The Scientific Basis of National Progress

The Scientific Basis of National Progress
Including that of Morality

G. Gore

First published in 1970 by Frank Cass and Company Limited

This edition first published in 2018 by Routledge
2 Park Square, Milton Park, Abingdon, Oxon, OX14 4RN
and by Routledge
52 Vanderbilt Avenue, New York, NY 10017, USA

Routledge is an imprint of the Taylor & Francis Group, an informa business

Publisher's Note
The publisher has gone to great lengths to ensure the quality of this reprint but
points out that some imperfections in the original copies may be apparent.

Disclaimer
The publisher has made every effort to trace copyright holders and welcomes
correspondence from those they have been unable to contact.

A Library of Congress record exists under ISBN:

ISBN 13: 978-0-367-14785-3 (hbk)
ISBN 13: 978-0-367-14786-0 (pbk)
ISBN 13: 978-0-429-05341-2 (ebk)

THE SOCIAL HISTORY OF SCIENCE
No. 4

General Editor: Dr. ROY M. MACLEOD
Reader in History and Social Studies of Science
at the University of Sussex

THE SCIENTIFIC BASIS

OF

NATIONAL PROGRESS

THE SCIENTIFIC BASIS

OF

NATIONAL PROGRESS

INCLUDING THAT OF MORALITY

BY

G. GORE

WITH A NEW INDEX

FRANK CASS & CO. LTD.

1970

Published by

FRANK CASS AND COMPANY LIMITED

67 Great Russell Street, London WC1B 3BT

| First edition | 1882 |
| New impression | 1970 |

ISBN 0 7146 2407 1

Printed in Great Britain by Clarke, Doble & Brendon Ltd.
Plymouth and London

THE SCIENTIFIC BASIS

OF

NATIONAL PROGRESS,

INCLUDING THAT OF MORALITY.

BY

G. GORE, L.L.D., F.R.S.,

Author of " The Art of Scientific Discovery ;"
" The Principles and Practice of Electro-deposition ;"
" The Art of Electro-metallurgy ;" &c.

NATIONS ADVANCE BY NEW KNOWLEDGE.

WILLIAMS AND NORGATE,

14, HENRIETTA STREET, COVENT GARDEN, LONDON;
AND 20, SOUTH FREDERICK STREET, EDINBURGH.
1882.

To the President (the Rev. N. Watson, F.R.S.), the Vice Presidents, the Council and Members of the Birmingham Philosophical Society, I dedicate the following small treatise, in appreciation of the fact, that although only a young Society, they have certified in a substantial manner the views persistently advocated by me respecting the National importance of Scientific Investigation, and have shown so intelligent an example of devotion to public welfare by establishing a Fund for the Endowment of original Scientific Research.

GEORGE GORE.

The Institute of Scientific Research,
Birmingham.

THE SCIENTIFIC BASIS OF NATIONAL PROGRESS.

CONTENTS.

THE SCIENTIFIC BASIS OF NATIONAL PROGRESS.

PREFACE.

As there exists at the present time in this country a con-siderable degree of uneasiness in the public mind respecting our ability to maintain our position in the race of progress, and as our future success as a nation depends largely upon science, it is desirable to call attention to the great public importance of *new* scientific knowledge, and to the means of promoting its development.

Although the illustrations given in this book of the im-portance of such knowledge to mankind, constitute but a small fraction of the number which might be adduced, they are sufficient to show that by the neglect of scientific investi-gation, we are sacrificing our welfare as a nation to an enormous extent.

The greatest obstacle to the discovery of new knowledge in this country, lies in a wide spread ignorance of the de-pendence of human welfare upon scientific research. I propose therefore to show in a brief manner, that the essential starting-point of human progress, lies in scientific discovery; also that new truths are evolved by original re-search made in accordance with scientific methods; and to illustrate these statements by examples; also to point out how such research can be encouraged.

Preface.

The book is divided into four chapters, viz. : 1st. The Scientific basis of Material progress : 2nd. The Scientific basis of Mental and Moral progress : 3rd. New truth and its relation to Human progress : and 4th. The Promotion of original Scientific Research. As the object of the book is only to call attention to the vast importance of *new* truth, as as a fundamental source of advance, and how to promote the discovery of it, the essay is written as briefly as possible, and is not offered in any sense as a complete exposition of the subject, especially the section relating to the Scientific basis of Morality.

The leading idea of the Book is that present knowledge only enables us to maintain our present state, that national *progress* is the result of *new* ideas, and that the chief source of new ideas is original research.* That as *advance* has its origin in *new* knowledge ; unless new discoveries are made, new inventions and improvements must sooner or later cease. Another prominent idea is, that truth is essentially the same in all divisions of knowledge, and that the mental powers and processes employed in detecting it are the same in all subjects.

For reasons stated in the text, the influence of scientific discovery upon mental and moral progress are treated together. Notwithstanding the far greater importance of the mental and moral advantages of new truths, the book treats chiefly of the pecuniary and material gains to mankind ; mainly because the latter are more easily understood and appreciated, the chapter however on " The Scientific Basis of Mental and Moral progress," indicates in a very brief and imperfect manner, the vast importance of new scientific knowledge to mankind, as a source of mental and moral advancement.

* See p.p. 165 to 167.

Preface.

The chief object of this book is to disseminate more correct ideas respecting the importance of *new* positive knowledge, and the duties of society in relation to it; and a further object is to assist in maintaining Birmingham in the front rank of intellectual, social and moral advance, in accordance with its motto "Forward."

CHAPTER I.

THE SCIENTIFIC BASIS OF MATERIAL PROGRESS.

DURING the last one hundred years this nation has advanced with unexampled speed. More wealth has been accumulated by Englishmen since the commencement of the present century, than in all preceding time since the period of Julius Cæsar; one of the causes of this has been the discovery of new truths of science, and their subservience to useful purposes by means of invention. The great manufacturing success of this country has been largely due to those applications of science, which have enabled us to utilise our abundant stores of coal and iron-ore, in steam engines, machinery, and a multitude of mechanical, physical, and chemical processes; also to the discovery of electro-magnetism and its application in the electric-telegraph, etc. And had it not been for these and other adaptations of scientific knowledge, we should have competed in vain with the cheaper labour and longer days of toil of continental nations. Other great causes, such as our insular position, suitable climate, freedom, geographical position, etc., etc. have, however, also contributed to the result. Commerce also in its turn has done vast things for mankind.

The purely scientific knowledge we possess was discovered almost entirely by means of original research, and to only a small extent by persons engaged in industrial occupations. Probably not two per cent. of all the important discoveries in pure science were made in manufactories; the scientific experiments which are made in such establishments are usually of the nature of invention, not of discovery, and are not often published, because it is a usual object with men of business to retain as much as possible of the pecuniary benefit of their labours to themselves. Whilst it is the object of a business man to monopolise special knowledge; that of the scientific man is to diffuse it, in order that all mankind may be benefited and helped to improve.

Discoveries in science are, however, occasionally made by practical men engaged in technical employments. The hydro-electric machine originated in this way, a man at Newcastle was attending to a steam boiler, and found that he received electric shocks when he touched the boiler. This circumstance was investigated by his employer, Mr. Armstrong, a scientific man, and led him to construct the hydro-electric machine. The accumulation of electricity in submarine telegraph cables was first observed at the Gutta-Percha Company's works

London. It was noticed on testing a cable by means of a voltaic battery (the cable being submerged in water) that discharges of electricity flowed from the cable after the battery was removed; this circumstance was investigated by Faraday, and led to improvements in submarine telegraphy. In each of these instances the same general method as that used by scientific discoverers was however employed, viz., new experiments were made (though not intentionally) by putting matter and its forces under new conditions, and new results were observed.

Nearly all great modern scientific discoveries have been made by teachers of science and others, who spend a large portion of their lives in experimental investigation, searching for new truths, and not by persons who have hit upon them by accident. The greatest discoveries in physics and chemistry in modern times, were made chiefly by such men as Newton, Cavendish, Scheele, Priestley, Oersted, Volta, Davy and Faraday: all great workers in science.

It is either by observing matter and its forces under new conditions or from a new aspect, that nearly all discoveries are made; thus Priestley placed some oxide of mercury in an inverted glass vessel, and heated it by means of the Sun's rays and a lens, and discovered Oxygen. This substance was nearly discovered by Eck de Sulsbach three hundred years before; he heated six pounds of an amalgam of silver and mercury, and converted the latter metal into a red oxide like cinnabar, and he remarked, " a spirit is united with the metal, and what proves it is this, that this artificial cinnabar submitted to distillation, disengages that spirit." The " spirit" was evidently oxygen.

Some discoveries are made by observing the phenomena of bodies placed under special conditions by those operations of nature over which we have little or no control. All our knowledge of Astronomy, and much of that of geology and physiology, was acquired in this way.

Nearly all modern discoveries of importance in physics or chemistry require long and difficult investigations to be made in order to completely establish their truth. When Crookes discovered Thallium, he saw the first sign of its existence in a momentary flash of green light in a spectroscope, but he had to expend upon the subject several years of most difficult labour, and a considerable sum of money, in order to prove the correctness of his suspicion that he had discovered a new metal. M. Lecocq de Boisbaudran discovered the metal Gallium and Bunsen discovered Rubidium and Caesium in a similar manner.

Discoveries in science, are usually made, not by trying to obtain some valuable commercial or technical result, but by making new, reliable, and systematic investigations. By investigating the chemical action of electricity upon saline bodies,

Sir Humphrey Davy isolated sodium and magnesium, which has led to the establishment at Patricroft near Manchester, of the manufactures of those metals. By the abstract researches of Hofmann and others upon Coal-tar, many new compounds were discovered, and the extremely profitable manufacture of the splendid coal-tar dyes was originated.

Scientific discovery is the most valuable in its ultimate practical results when it is pursued from a love of truth as the ruling motive, and any attempt to make it more directly and quickly remunerative by trying to direct it to immediately practical objects, decreases the importance of its results, diminishes the spirit of inquiry, and sooner or later reduces it to the character of invention. The greatest practical realities of this age had their origin in a search after important truths entirely irrespective of what utilities they might lead to.

I do not intend by these remarks to imply that any new trades or improvements in manufactures have been or can be effected without the labours of inventors and practical men, but that there should be a more judicious division of labour : one man to discover new truths, another to put them into the form of practical inventions, and the business man to work them ; because it is proved by experience, that in nearly all cases these different kinds of labour require men of widely different habits of mind, and that the faculties of discovery, invention, and practical working are very rarely united in one man.

Scientific investigations however, made in a manufactory, for the purpose of ascertaining the various sources of loss of materials, the circumstances which affect the amount or quality of the product ; or made with the object of substituting cheaper or more suitable materials, or for varying their proportions, or for many other kindred objects, have in many cases resulted in great benefit to the manufacturer, and have formed the basis of successful patents. Some of the large brewers, chemical manufacturers, candle companies, and many others, constantly employ scientific men in this way to examine their materials, processes and products, and keep them acquainted with the progress of discovery and invention in relation to their own particular trades.

No art or manufacture is so perfect as to be exempt from the influence of discovery and invention, and no man can produce so perfect an article but that, by the aid of science, a better may be produced. Science and trade are mutually dependent, without the assistance of science, trade would be unable to supply our daily increasing wants, and without the pecuniary support of trade, science would languish and decay.

" As long as arts and manufactures are left to be directed and improved by simple experience, their progress is extremely slow, but directly scientific knowledge is successfully applied to them, they bound forward with astonishing speed." Look at the art of taking portraits; for hundreds of years it remained entirely in the hands of oil and water-colour painters with but little progress in rapidity of production, but directly science was applied to it in the form of photography, its advance in this respect became amazing. Fifty years ago photography was almost unknown, but immediately Messrs. Daguerre and Talbot, in 1844, made known their processes, the new art began to advance, and so rapid has been its progress, that at the present time many thousand persons are employed in its exercise, and millions of portraits have been taken with an accuracy and at a cost quite beyond the reach of the old method.

Many persons hardly know the difference between science and art; a still greater number cannot readily distinguish between a concrete science and a pure one; and nearly all persons confound discovery with invention. A science may be conveniently defined as a collection of facts and general principles which are to be learned; an art as a collection of rules which are to be followed:—Art therefore is applied science; and every art also has a basis in science, whether that basis has been discovered or not. Scientific principles underlie not only manufacturing processes, but also sculpture, music, poetry and painting.

Discoveries differ also from inventions: a scientific discovery is a newly found truth in science, which in the great majority of cases is not in the form of applied knowledge. An invention is usually a combination and application to some desired purpose, of scientific truths which have been previously discovered. When Oersted first observed a magnetic needle move by means of a current of electricity, he made a scientific discovery; but when Wheatstone and Cooke applied Oersted's discovery in their telegraph from Paddington to Slough, they made an invention. The success of the electro-plating process was dependent upon knowledge previously discovered. Mr. Wright, a surgeon in Birmingham, was led to the invention of the use of cyanide of potassium in electro-plating and gilding, by reading in Scheele's "Chemical Essay" (p.p. 405 and 406), that "if after these calces" (*i.e.*, the cyanides of gold and silver) "have been precipitated, a sufficient quantity of precipitating liquor be added, in order to redissolve them, the solution remains clear in the open air, and in this state the ærial acid" (*i.e.*, carbonic acid of the air) " does not reprecipitate the metallic calx."

Immediately a discovery is effected it is made public, and is afterwards incorporated in the ordinary text books of science, ready for the use of inventors; and in this way such books have become filled with valuable knowledge acquired by researches in past times. All this knowledge (which has cost millions of pounds and a vast amount of intellect and labour) has been given by its discoverers freely to the nation. Some idea of the number of scientific researches which have been made since the year 1800, may be obtained from the fact, that a mere list of their titles, with the names of the authors, occupies eight large quarto volumes, of about one thousand pages each, compiled and published at a cost of about ten thousand pounds, by the British Government and the Royal Society.

In discovery we search for new phenomena, their causes and relations; in invention we seek to produce new effects, or to produce known effects in an improved manner. The objects of the scientific discoverer are, new truth and greater accuracy; whereas those of the inventor, are increased usefulness and economy of results. The ancients classed inventors with the gods, because they considered them great benefactors to the human race. Discoverers may properly be viewed as priests and prophets of truth, because they both reveal new knowledge to mankind, and predict with certainty coming events.

A man cannot usually invent an improvement unless he possesses scientific knowledge, and, for that knowledge he must in nearly all cases resort to a scientific book or teacher. The great practical value of new scientific knowledge is proved by the fact, that when scientific discoveries are published, there are numerous inventors and practical men, who immediately endeavour to apply them to useful purposes. Since the first application of coal-tar to the production of dyes, every discovery in that branch of chemistry has been closely watched for a similar purpose.

A complete account of the growth and development of scientific discoveries and inventions would form an extensive history, and would include numerous instances of experiments attended by results which, sooner or later, affected all mankind. Take that of phosphorus, for example. The first evidence of the existence of that substance was obtained by the Saracens in the eighth century. Achild Bechil distilled a powdered mixture of charcoal, clay, lime, and dried extract of urine, and obtained a substance which shone in the dark "like a good moon;" that substance was phosphorus. The discovery contained in the results of that little dirty and stinking experiment was the germ or seed of all the subsequent developments and applications of phosphorus. About the year 1669 Bechil's experiment was further developed by Brandt, a merchant of Hamburg, and the publication of the wonderful properties of the substance produced a great sensation in his fellow-citizens. "There was then cried nothing but triumph and victory among the chymists. Those good people erected already in their thoughts so many hospitals and poor-houses that no beggar should more molest any man in the streets, made great legacies, and pious causes, and what not else." "Besides, the other alchymists did encourage him yet more, and desisted not to make him believe how this was the same fiery ghost of

Moses that in the beginning moved upon the water, yea, his splendid shining face : the fiery pillar in the desert, that secret fire of the altar wherewith Moses burned the golden calf before he strewed it upon the fire and made it potable."

The experiment of Brandt was repeated by Kunckel before the courts of Saxony and Brandenburg, although it was not a very delicate or agreeable exhibition, "because the anctuous and daubing oyliness was not yet accurately separated from it, and without doubt it was very stinking." Brandt's process was further developed by Boyle, and published in the Philosophical Transactions of the Royal Society, in the year 1692-3 ; and phosphorus was afterwards obtained in larger quantity and in a purer state by Hanckwitz, a chemist in Southampton Street, Strand, and sold by him at three pounds sterling per ounce. Its price at present is less than three shillings per pound.

Margraaf, Fourcroy, Vauquelin, and Dr. Slare also extended our knowwledge of the substance ; Gahn, in 1769, made the important discovery of phosphorus in bones, and Scheele immediately devised the process now in use by our manufacturers for extracting it from that substance. The commencement of the use of phosphorus for the purpose of getting a light occurred about the year 1803, but it was not until the year 1833 that the invention of phosphorus matches became commercially successful. The use of such matches is now universal, and it has been estimated that the daily consumption of them in Great Britain alone amounts to two hundred and fifty millions, or more than eight matches per day for each individual in the kingdom.

"There is nothing on the Earth so small that it may not produce great things." The most abstract and apparently trivial experiments in original research have in some cases led to inventions and results of national and even world-wide importance. The contractions of a frog's leg in the experiments of Galvani, and the movements of a magnetic needle in those of Oersted, have already led to the expenditure of hundreds of millions of pounds in laying telegraph wires all over the earth, and to an immense extension of international intercourse. But the original experiment of Oersted was not discovered without labour, it was only arrived at after many years of research.

The saying that "all great things have had small beginnings," is true, not only of electric telegraphs, but also of the great trade of electro-plating, and of the magneto-electric machine which is now largely used instead of the voltaic battery. After Volta had made his small and apparently unimportant experiments on the electricity produced by metals and liquids, various persons tried the effect of that electricity upon metallic solutions. Brugnatelli, in 1805, found that two silver medals became gilded in a solution of gold by passing the electricity through them. Mr. Henry Bessemer, in 1834, coated various lead ornaments with copper by using a solution of copper in a similar manner. And in 1836 Mr. De la Rue found that copies might be taken in copper of engraved copper-plates by the electro-depositing process. Faraday discovered magneto-electricity in the year 1831, by rotating a disc of copper between the poles of a magnet, and he has stated that the first successful result he obtained was so small that he could hardly detect it. This simple experiment was the origin of the magneto-electric machine, and many of these machines are now used for producing the electric light, and for depositing nickel, copper, silver, and gold, instead of by the voltaic battery. These, and other engines, thermic, magnetic, electric, &c., will probably, ere long, be constructed on as large a scale, and as many in number, as the present steam engine.

The discovery in olden times of the attractive properties of a fragment of iron ore, was the basis of the invention of the mariner's compass, which greatly improved navigation, and led to nearly all the chief maritime discoveries which have since been made. The sciences of magnetism and geometry, form the basis of the art of navigation, and have thus made our great foreign commerce possible. The discovery of magnetism enabled sailing vessels to venture freely out of sight of land, and to traverse the wide ocean with even greater safety than to sail near the shore. By its means Columbus crossed the Atlantic Ocean and discovered America. By its means also, Vasco de Gama sailed round the Cape of Good Hope and discovered a new route to India ; and in the year 1500, another Portuguese Captain, Cabral, was driven across the Atlantic, discovered Brazil, and was enabled by the aid of the magnet, to send back a ship to Lisbon with news of the discovery. By its assistance also Magellan discovered Patagonia and the South Pacific Ocean ; and by the completion of that voyage the Earth was first circumnavigated and proved to be a globe.

The geographical discoveries of the Portuguese, made by means of the magnet, produced great national results ; they profoundly changed the balance of power and wealth among European nations, by changing the direction of navigation and of the great streams of commerce between Europe and the East. They gave a mortal blow to Italy and the cities of the Mediterranean, by transferring Eastern commerce to Spain and Portugal : and Egypt ceased to be the greatest route of commerce from Europe to India.

A singular contract relating to geographical research was made in the fifteenth century, between King Alphonso, of Portugal, and Ferdinand Gomez, of Lisbon, by which the latter engaged to navigate a ship and explore the coast of Africa, and to discover not less than three hundred miles of coast every year, the measurement to be made from Sierra Leone.

Scientific discovery has in all ages been a most powerful agent of civilization and human progress. The discovery of the black liquid which a solution of nutgalls produces when mixed with green vitriol, led to the invention of writing ink; and a knowledge of the properties of ink and paper prepared the way for the invention of printing, by means of which truth and learning have spread all over the earth.

The apparently insignificant property possessed by amber, of attracting feathers immediately after it has been rubbed, was known twenty-four hundred years ago, and afterwards led to the discovery of electricity. In later times, Dr. Franklin, by means of a kite, charged a bottle with lightning, examined it, and proved lightning and electricity to be identical. This knowledge, joined to the further discovery, that electricity would pass freely through metals, led to the modern invention of the lightning conductor, by means of which all our great buildings, ships, lighthouses, arsenals, and powder magazines are protected from lightning.

"Coming events cast their shadows before them:" the discovery of the instant transmission of electricity along wires by Stephen Gray and Wheeler, about the year 1729, fore-shadowed the invention of the electric telegraph. About the year 1819, Oersted, a Danish philosopher, after fifteen years of study and experiment, to ascertain the relation of electricity to magnetism, discovered that if a freely suspended magnetic needle was supported parallel and near to a wire, and an electric current then passed through the wire, the needle moved and placed itself at right angles to the current. This discovery, coupled with the previous one of the electric conductivity of metals, formed the indispensable basis of all our electric telegraphs.

Original research is very productive of new industries and inventions. The discoveries made by Volta, Faraday, and many other investigators, have led to the process of electro-plating, the use of electric lights for lighthouses, and for ocean steamships, and the great system of telegraphs. Those of Davy, Wedgwood, and others, respecting the action of light upon salts of silver, have resulted in the modern processes of photography, which are now in use almost everywhere. The discovery of zinc, by Paracelsus, has been followed by the use of that metal in galvanic batteries, and the great use of "galvanized" iron for telegraph wires, for roofing, and many other purposes. The discovery of nickel, by Cronstedt, has led to the great modern use of that metal in electro-plating, and to that of German silver in the construction of electro-plated and other articles. The discovery of chlorine, by Scheele, formed the basis of nearly all our modern processes of bleaching cotton and other fabrics. The discovery of gun-cotton and nitro-glycerine has led to the use of those substances in blasting rocks and in warfare. The discovery of oxygen, by Priestley, has enabled us to understand and improve in a great number of ways the numerous manufacturing, agricultural, and other processes in which that substance operates. Priestley made many experiments also on the absorption of gases by water, and proposed the resulting liquids as beverages; and those apparently trifling experiments have since expanded into the large manufactures of aërated waters. The discoveries of gutta-percha and india-rubber were indispensible to the great applications of those substances in telegraph cables, and in a multitude of useful articles. The discovery of chloroform and anæsthetics has led to their use for the purpose of alleviating human suffering. The discovery, by Sir Isaac Newton, of the decomposition of light by means of a prism, has led in recent times to the invention of the spectroscope; to the use of that instrument in the Bessemer steel process; to the discovery of a number of new metals, thallium, rubidium, cæsium, indium, and several others, and to the most wonderful discovery of the composition of the Sun and distant heavenly bodies.

Even the invention of the steam-engine was partly a consequence of previous researches made by scientific discoverers. Watt, himself, stated in his pamphlet, entitled "A plain Story," that he could not have perfected his engine had not Dr. Black and others previously discovered what amount of heat was rendered latent by the conversion of water into steam. "Each mechanical advance in the steam-engine has been preceded by and the result of the

discovery of some physical law or property of steam."
" The first step in the invention of the steam-engine
was the experimental research and the discoveries of
the properties of steam by Hooke, Boyle, and
Papin."* Had not the steam-engine been developed,
it is clear that railways, steamships, machinery, and
all the other numerous uses to which that instrument
is now applied, would have been almost unknown.
The introduction of the steam-engine enabled abandoned Cornish mines to be relieved of water, and to
be worked to much greater depths.   The discoveries
of nitric acid, hydrochloric acid, oil of vitriol, and
washing soda, by the alchemists and early chemists
in their researches, led to the erection of the
numerous great manufactories of those substances
which now exist in England and in other civilized
countries. There is probably not an art, manufacture,
or process, which is not largely due to scientific discovery, and if we trace them back to their source
we nearly always find them originate in scientific
research.

So far has scientific discovery, and its practical
applications to human benefit by invention, now
progressed, that every one considers this to be, *par
excellence*, the scientific age. And as discovery and
invention continue to progress with accelerated speed,
we are encouraged to hope, not only that scientific
principles will ultimately be universally recognised as

* Essays and Addresses, Owen's College, 1874, pp. 172-182.

the regulators of all technical industry, but also as
a fundamental basis of morality.*

"It is true that some processes of manufacture have
not been consequences of abstract scientific discovery
—that they originally resulted from alterations made
in the rudest appliances, and that they have been
directed and improved by the results of simple
experience. For ages past we derived the benefit of
scientific principles without a knowledge of their
existence. We trod in the beaten paths of experience ignorant of the truth that we were acting in
unison with fixed and certain laws. Numerous arts
and processes were in extensive operation long before
the principles involved in them were at all understood. The arts of enamelling and of iron smelting
were known hundreds of years before we were
acquainted with the principles of chemistry. In
some rare instances also the recorded results of daily
experience in practical matters, tabulated and
studied, have ultimately led to the discovery of
scientific laws; but all this is merely the making
use of our ordinary experience for the advancement
of knowledge, instead of making *special experiments*
for the purpose."

Many of our processes and manufactures, those of
glass and copper for example, are of such great anti-
quity, it is impossible to ascertain with certainty the
special circumstances under which they originated;
but after we have fully considered the ways in which

* See Chapter 2, Section B.

various modern trades and manufactures have first arisen, we shall come to the conclusion that all manufactures and improvements in manufacturing processes, must have been first produced by the same general means, viz., new observations, although the special circumstances connected with the origin of each were different.

Let us consider German-silver and its manufacture. That substance is an alloy of copper, zinc, and nickel; it owes its peculiar whiteness or "silver-like" appearance to the latter metal, and cannot be made without it; it is certain, therefore, that by whatever means that metal or the alloy was discovered, the discovery was the origin of the German-silver manufacture, and was essential to all manufactures, processes, or appliances in which German-silver, nickel, or any of its compounds are used. Nickel was discovered by Cronstedt during the year 1751, and its compounds were chiefly investigated by English and foreign chemists. Cronstedt found it as a peculiar metal in the mineral called kupfernickel, whilst chemically examining the properties of that substance. The general method by which he discovered it was careful experiment, observation, and study of the properties of matter.

It is stated that the Chinese and other nations made alloys of nickel long before nickel itself was known to be a distinct metal; they had found, by experiment, that when ores of copper and zinc were mixed with a particular kind of mineral and smelted, a white alloy was obtained; but this also proves the general statement. already made, that the German-silver manufacture was originated by means of new observations. It was by a more skilful, but similar mode of procedure that Cronstedt isolated the metal itself, and thus laid a definite basis of improvements in the manufacture of its alloys.

No art is probably more antique, or remained longer exempt from the influence of science, than that of match making and obtaining a light. Many adult persons can remember the primitive and old-fashioned tinder-box, which had passed, with its flint and steel, from one generation to another without any material improvement. Phosphorus, it is true, was definitely discovered at least as early as the year 1669, but it was not applied to match making till about 1833. Since then the progress of invention has been so rapid that there are now numerous manufactories which produce many millions per day of phosphorus matches; for instance, those of M. Pollak, at Vienna, and of M. Fürth, in Bohemia, consume together more than 20 tons of phosphorus annually, and give employment to about 6,000 persons, and as one pound of phosphorus suffices for about one million German matches (or 600,000 English ones), those two makers alone produce the astonishing number of 44,800 millions of matches yearly.

Judging by means of the experience already acquired, we cannot reasonably expect that discoveries fraught with such momentous consequences as those of magnetism or of galvanism and electro-magnetism, will be made very often. The progress of scientific discovery is gradual; we have at present but mere glimpses of the new world of truth which is being revealed to us by means of research; we are only at the very commencement of a knowledge of the inherent properties of matter and its forces, and consequently the methods we employ to utilize them are extremely imperfect. Matter has a general property of subdividing and transmuting forces; if we apply one force to a substance or machine, it produces many effects, not only those we want, but those also we do not want; when we heat a piece of iron, the heat produces a number of changes, mechanical, electric, magnetic, and chemical, and it is partly by means of what is termed the "internal resistance" of bodies that these effects are produced, and we know but little of that property. The ex-plosive action in a gas engine produces not only the mechanical force we desire, but also a quantity of heat we do not want, and at a cost of a portion of the gas. In a similar manner, in the steam-engine the largest portion of the heat of the coal is con-verted into forces which are lost; a large amount of it is uselessly expended in warming the machine itself and the surrounding atmosphere; much also is lost by friction.

That "knowledge is power" is an old maxim, but that *new knowledge is new power* is a new maxim which scientific discovery has impressed upon us. By means of discoveries we have acquired new powers ; by those of electricity we have acquired the ability of conversing with each other at unlimited distances, and by means of those in optics we are enabled to analyse the composition, and perceive some of the physical changes of the most distant heavenly bodies. As our ignorance is probably much greater than our knowledge, more inventions also, and extensions of human power, must ultimately result from discovering new qualities of bodies, than by applying to useful purposes their already known properties.

Experience in science has already shown that it is by means of invention based upon *new* discoveries that the greatest utilities are obtained, rather than by the exercise of invention upon knowledge acquired long ago. The information obtained by research in former times has been largely exhausted for the purposes of invention by modern inventors, and what we very greatly require now is *new* knowledge. Experience in science also leads us to believe that the extent of possible discovery is as boundless as Nature, and that an immense amount of new knowledge may yet be discovered. Every discoverer of repute could supply a copious list of investigations yet to be made.

An infinite number of questions in pure science remain to be decided by means of research. Is Electricity decomposible like radiant heat or light ?

Are the "elementary substances" really compound bodies? Are they all compounds of Hydrogen? Are they all decomposed by very high temperatures, as compound substances are "disassociated" by less elevated temperatures? Under what conditions is Fluorine isolated? Do gases transmit heat by conduction? Under what circumstances is Light converted into Electricity? and into Magnetism? What is the actual size of an atom of Hydrogen? Does Light (without heat) expand bodies? What is the actual molecular arrangement of the atoms of Hydrogen at 60 Fahrenheit? What is the cause of the absence of metalloids in the Sun? What are the properties of Fluorine? What is the vapour density of Cæsium? Under what circumstances is heat wholly converted into mechanical power? &c., &c. All these discoveries when made, will probably, sooner or later, be productive of practical benefits to mankind.

Nearly ever manufacturer in this country is deriving, from scientific discoveries, advantages for which there has been little or no payment made to the discoverers. The makers of coal-tar-dyes, and dyers of wool and silk, are using Mitscherlich's discovery of nitro-benzine. Manufacturers of picric acid and "French purple" have enjoyed the fruits of the labours of Dr. Stenhouse. Makers of chlorate of potash and cyanide of potassium are profiting largely by the discoveries of Scheele, Gay-Lussac, and others. All the percussion cap makers

are indebted to Howard and Brugnatelli for fulmi-
nating silver. Railway-contractors, quarry-pro-
prietors, and others, use nitro-glycerine discovered
by Sobrero. Iron smelters are benefiting by the
discovery of Bunsen, that 42 per cent. of the heat of
the fuel was lost as combustible gases—these gases
are now utilized. Telegraphists and electro-platers
are also indebted to him for his voltaic battery. The
producers of metallic magnesium owe the origin of
their process to him as being the first to convert it
into wire ·and make known its great light giving
power. Multitudes of persons now use his well-
known "Bunsen's burner" for heating, cooking, and
other operations. The various telegraph companies,
copper smelters, and makers of copper telegraph
wire, are using Dr. Matthiessen's discovery of the
influence of impurities on the electric conducting
power of copper. Phosphorus-makers are reaping
the reward of the labours of Gahn and Scheele. The
makers of electro-plate and German silver are
deriving profits from the labours of Faraday, who
investigated electrolysis; of Gay Lussac, who discovered cyanogen; and of Cronstedt, who discovered
nickel. Makers of Bessemer steel enjoy advantages
derived from the spectrum discoveries of Kirchoff.
Iron and copper smelters, metallurgists in general,
dyers, calico printers, bleachers, brewers, makers of
vinegar, red lead, varnishes, colours, soaps, green vitriol, phosphorus, oil-of-vitriol, and many others, are
deriving benefit from the discoveries of Priestley and

Scheele. Physicians and their patients are receiving the reward of the labours of Soubeiran, Liebig, and Dumas, in the discovery of chloroform; of the researches of Fourcroy, Vauquelin, Pelletier, and others, in the discovery of quinine; and of many other chemists who discovered numerous remedial substances. By means of the discoveries of Oersted and others, embodied in the telegraph, manufacturers are enabled to anticipate the state of the markets and of the weather, and editors are enabled to obtain the earliest news.

Suppose that Gay Lussac, in 1815, had not discovered cyanide of potassium, and that it had never been discovered, it is highly probable that the manufacturing returns of Birmingham and Sheffield would be much less in amount at the present time than they are, simply because there is no other known substance with which the electro-plating of base metals with gold and silver can be satisfactorily effected. Or suppose that sal-ammoniac, chloride of zinc, or other soldering agents had not been discovered, the extensive and so-called "galvanizing" process could not have been effected, because without those substances the iron articles immersed in the melted zinc would not have received an adhesive metallic coating.

On the other hand, science has in various cases rendered obsolete some manufactures and superseded old customs, comforts and conveniences. We have ceased, or almost so, to use tinder-boxes, snuffers,

sulphur matches, rush-lights, tallow candles, sedan
chairs, stage coaches, the ancient water-bucket and
well, and even the comparatively modern pump;
coal fires also are gradually being superseded by fires
of gas, and articles formed of solid silver are now
being replaced by those of electro-plate; canals
have also to some extent been supplanted by railways. But in all these cases science has supplied us
either with something better or more suited to our
present wants.

The great pecuniary benefits arising from the
applications of science are generally reaped in the
first instance by the great manufacturers, agricul-
turists, merchants, and capitalists. Countless for-
tunes have been made by means of processes and
manufactures based upon scientific discovery. The
pecuniary benefits of calico printing, bleaching, dye-
ing; of the great manufactures of cotton, iron,
pottery, beer, sugar, glass, spirits, vinegar, gutta-
percha, india-rubber, gun cotton, the numerous
metals, machinery, electro-plate, washing soda,
German silver, brass, phosphorus, manures, the
common acids, numerous chemicals, and a multitude
of other substances and articles, have been extremely
great. More than eighteen hundred million pounds
of sulphuric acid alone are manufactured in Europe
yearly. The pecuniary advantages of the use of the
electric telegraph and railways to merchants, the
gains of capitalists by monies invested in railways,
telegraphs, steam-ships, cotton-mills, gas-works,

iron shipbuilding, engineering, and other great applications of science, have been enormous. The annual gas rental of London alone amounts to more than two millions sterling; and even in Birmingham the produce of gas is more than twenty-five hundred millions of cubic feet yearly. The amount of capital expended in the construction of railways only in this country, has been estimated at more than seven hundred millions of pounds, and the total receipts upon British railways has reached forty-three millions per annum. In the year 1875 our railways carried 200 million tons of goods, and consumed ten million tons of coal; the Great Northern Railway alone consumes 5,000 tons of coal each week. In the year 1877 there existed in the entire world about 198,000 miles of railway, the whole having been constructed since the year 1825. In the year 1880 six hundred millions of journeys were made by passengers on British railways; and the stock of those railways included 13,174 locomotives; 369,694 waggons, 28,717 passenger carriages, and 22,712 other vehicles. The London and North-Western Railway Company alone possessed, in the year 1873, no less than 1,900 locomotive engines, each of a value of nearly two thousand pounds; 4,000 carriages and 36,000 waggons; and it has been estimated by competent authorities, that there are in the world 200,000 steam-engines, having a total power of twelve million horses, or 100 million men. The number of cotton spindles on the whole Earth is estimated at about $71\frac{1}{4}$ millions. In the United States of America there are about five thousand telegraph stations, and 75,000 miles of line, which transmit yearly about 11,500,000 messages.—The telegrams of Great Britain number about one-fourth of a million per week. The world's telegrams during the year 1877 numbered nearly 130 milllons; and the world's letters about 3,300 millions, or $9\frac{1}{4}$ millions each day. Even the little phosphorus match is being manufactured and consumed at a rate estimated at more than ten thousand millions daily.

Much of the wealth of this country, resulting from science, has been very easily obtained by its pos-sessors. That acquired by means of our coal has especially been obtained without commensurate effort. The amount of that substance raised in Great Britain during the year 1876 was 734 millions of tons. To draw upon a great stock of that mineral is like drawing money from a bank, because coal, unlike any other abundant substance (except wood and petroline), contains in itself an immense store of energy, which is evolved as heat during combustion, and may be utilized. Each piece of coal contains sufficient energy to lift its own weight twenty-three hundred miles, but it costs only a small proportion of that power to extract and raise it from the mine. I do not mean by these remarks to imply that the wealth accruing from this great store of power in coal is derived chiefly by the owners of coal mines,

This acquisition of wealth without commensurate sacrifice is not an unqualified advantage; it constitutes a debt to nature, which upon the great principle of causation, and of equivalency of action and reaction, must sooner or later be repaid. Judging from the infallibility of the action of those laws, and the signs of the times, this nation is now beginning to repay in the form of emigration of trade to other lands, and of relatively less rapid national advance, the debt incurred by undue pecuniary success. An excess of money or power obtained without equivalent effort, fails to properly develop the intelligence of its possessors, and nations have been hastened to ruin in this way. Our great success in getting money has attracted many from the pursuit of knowledge, and our love of knowledge has not increased as fast as our wealth. The wealth of the upper classes has, by decoying from study undisciplined young men at our old Universities, kept down the general standard of scientific instruction throughout the country, and, by leading to neglect of scientific research, is now retarding our progress in arts, manufactures, commerce, and civilization. The consequent relative poverty of the working classes is also producing similar effects by retarding education, and contributing towards the great deficiency of skilled labour, of which our inventors, manufacturers, and others so strongly complain in the working of their scientific processes. Had a just share of the great amount of money, gained by the application of science to useful purposes, been applied to the payment and maintenance of scientific discoverers and inventors, as it should have been, the general standard of scientific education would have been higher, the poor would have had more employment and money, and the happiness and civilization of all would have been greater.

In a usual way the greatest pecuniary benefits, arising from science, sooner or later go to enrich the possessors of land. The demand created for coal, iron, lime, building-stone, and all the metals, by the industrial applications of science, has greatly increased the value of land under which those substances lie. The value of cultivated land has been everywhere increased by the discoveries of agricultural chemistry. Land has also been required for railways in nearly all parts of the kingdom, and has thereby been considerably raised in value. Discoveries produce inventions, inventions give rise to processes and manufactures, the employment of workmen and others, and the erection of workshops and dwellings, and these have rapidly increased the value of building ground. In Lancashire the value of such ground has been greatly increased by the inventions of the steam-engine and machinery, the discovery of chlorine, and their application to cotton manufacture. In all the great manufacturing districts, and in all the chief centres of industry, a similar result has occurred. Wherever a railway has been constructed, the value of land has also increased in

consequence of the increased facilities of communication. All these great additions to the value of
land are largely due to the unpaid labours of scientific discoverers, and it may be said that this nation
has largely gained its wealth, and is still living in a
great degree on the products of those labours.
Those great additions to the value of land are also
permanent, are continually increasing, and are
largely independent of any exertions on the part of
the owners. That many other influences, besides
that of science, have contributed to the development
of our manufacturing and commercial prosperity is
also true, but it would be foreign to the subject of
the present chapter to point them out.

It is a fallacious argument to say that scientific
discovery and increased value of land are only
remotely connected together, a cause as certainly
produces its effect, however many connections lie
between them, provided the connections are certain
—the number of links in a chain makes no difference in the transmission of motion from one
end of it to the other. Great causes are frequently
distant and wide-spread in their effects. Persons
in general ean easily understand that an acorn
planted in the ground will in the course of
time become an oak, because it is a palpable and
visible effect; but they cannot so readily perceive
that the benefits resulting from a knowledge of
science ramify through all our manufacturing, artistic,
and commercial occupations, our social and moral

relations, and our every-day life, not because the dependence of our welfare upon science is less real, but partly because the connection between the two is less understood.

Not only has science benefited manufacturers, but also operatives, because the extension of science to manufacturing purposes has compelled them to make themselves acquainted with intellectual subjects. "Instead of remaining mere machines, mechanically performing the work set before them, they are obliged to exercise the faculties of observation and judgment in watching the results and directing the action of mechanical, physical, and chemical powers. Instead of following the blind path of experience, using unknown forces to accomplish some definite result, they pursue their labours with the aid of known and certain laws." It is true that in many cases artisans who have acquired a little knowledge of science have thereby been rendered conceited and unfit for their special employment, and this has made many manufacturers object to technical scientific education for their servants; but this would not be so much the case if scientific knowledge were more generally and equally diffused. Arguments are not unfrequently adduced to support the opinion that ignorance has its advantages; but, however great the advantages of ignorance may be, those of knowledge are greater.

In consequence of the labours of scientific dis-coverers and inventors, the progress of science is such that in a very few years a knowledge of it will be indispensible to all persons engaged in superintending or carrying out manufacturing operations, and in all arts, occupations and appointments in which man is dealing with matter. Science is fast penetrating into all our manufactures and occupations, and "those who are unscientific will have much less employment and will be left behind in the race of life." England also will be compelled, by the necessities of human progress and the advance of foreign intellect, to determine and recognize the proper value of scientific research as a basis of progress. National superiority can only be maintained by being first in the race, and not by buying inventions of other nations.

The philosophy of matter is the foundation of all manufacturing arts and artistic processes; technical education, or the relation of science to manufactures, &c., can only be properly imparted upon the basis of a sufficient knowledge of theoretical science. Science tends to abbreviate mental and bodily labour. The use of our reason saves us the labour of using our senses, because it enables us to know that under certain conditions a certain effect must occur. The use of our reason and senses also saves us using our hands.

The properties of a single substance are so numerous that if a workman was to thoroughly study the whole of them, he would become a scientific authority in the subjects of heat, light, electricity, magnetism, and chemistry. A blacksmith who knew all the physical and chemical properties and rela-tions of iron and steel would be quite a scientific philosopher.

No man has more occasion to bless the introduc-tion of the steam-engine, machinery, the galvanic battery, and science in general, than the working mechanic, because it has mitigated his physical toil by giving him the duty of simply directing the labour instead of actually performing it; whilst it has deprived him of one kind of employment it has provided him with something better. But a few years ago the operatives in the silver-plating trade had to lay the silver on the articles with their hands, with the aid of a soldering iron; now they have simply to set their batteries in action and watch the electricity doing it for them. In a similar manner the working engineer at his metal-turning lathe has merely to direct the action of his tools whilst the steam-engine performs the heavy labour of turning.

There is not a man in this kingdom who has not derived some advantage, in one way or another, from scientific research. The advantages of gas light, electric light, rapid postal service and transmission of goods, railway travelling, steam-ships for naviga-tion, cotton apparel, photography, cheap pottery, improved medicine and surgery, telegraphic forecasts of weather, Australian preserved meats, &c., &c., have been reaped more or less by everyone, even the very paupers. Not only has travelling been considerably cheapened and immensely increased, but also rendered more safe:—in travelling by diligence in France the average number of persons injured was 1 to every 30,000 carried; and killed, 1 in every 335,000 ; but by railway, notwithstanding the average length of the journey has greatly increased, the former has been diminished to 1 in 580,000, and the latter to one in five millions ; safety in travelling by sea has also been greatly increased by means of improved lighthouses. By the rapid transmission of messages by telegraphs and of commodities by steam-ships and railways, the horrors of famine have been largely diminished ; the health of this nation has also been improved by greater variety of foods, and the increasing cost of meat has been restrained. It is well known that in periods of famine, the great loss of life has arisen, not from universal scarcity of food, but from the loss of time in ordering and conveying it. Whilst also the steam-engine has been the means of relieving hundreds of thousands of men from mere animal toil; it has, with the aid of the printing-press, supplied them with cheap daily intelligence.

Science has also proved itself to be a great source of employment, as well as wealth. By developing new processes it has given employment to whole armies of workmen in numerous arts, manufactures, and occupations. Some of those employments necessitating scientific training. About 300,000 persons are employed on railways alone in Great

Britain, besides those who were engaged in their construction; and in the postal department alone of the telegraph service of this country more than fifteen thousand operatives are employed. Chemical works also find employment for twenty-six thousand, and gasworks for ten thousand work people. The telegraphs of the United States of America alone, provide employment for about 7,000 persons; and the railways of the world employ about 1,900,000 men.

It may be objected that the extension of science in this country, instead of increasing employment for workmen has produced an opposite effect, by so increasing the production of goods by machinery, and by physical and chemical processes, that we have glutted the markets of the world in years gone by, and are now suffering the results of over-production. This is a very limited view of the case; over-production is only true of particular manufactures, and is a result of ill-directed commercial energy, to which manufacturing skill is only a servant. The objection also contains its own reply;—that it is certainly much greater to our advantage to have supplied other nations with manufactured commodities, than that other nations should have supplied us, as they would have done had they the manufacturing skill. At present, however, continental nations are gradually supplanting us in manufactures; and gradually supplying us with the goods which we formerly supplied them, and our fear is that this is largely a result of our neglect of science.

In many cases instead of superseding labour, science has changed its kind, or its mode of distribution;—in the case of steam-ships, instead of navigation being conducted entirely by nautical ability, it is partly effected by the skill of the engineer; conveyance of goods by road and canal has not been entirely supplanted, but partly supplemented by conveyance by railways. The diminution of labour which sometimes occurs in consequence of the progress of science is extremely small compared with its increase. The number of waggoners and horses now employed, merely to collect and deliver all the goods for railways, is actually much greater than the whole of those employed for conveying all the goods of the country before railways were constructed.

It would be altogether a false argument to say that the practical benefits derived from the labour of scientific discoverers by the different classes of the community are uncertain or imaginary, because the discoveries and the practical benefits are not in all cases immediately connected. We know that the consumers of tea in this country derive benefit from the grower of that herb in China through the hands of a series of intervening agents, as certainly as if they received the tea direct from his hands. Cause and effect are inseparable, and the remote effect of a series of connected causes is not less certain than the immediate ones.

It is a remarkable fact, that of the multitude of rich manufacturers, merchants, capitalists, and land-owners in this country, who have derived such great pecuniary benefits from original scientific research, there is scarcely one who has ever given to a scientific society, institution, or investigator, a single thousand pounds for the aid of pure research in experimental physics or chemistry;* the nearest approach to exceptions are a very few wealthy persons who have devoted themselver personally to scientific discovery. Manufacturers have willingly reaped the advantages of the labours of unpaid discoverers, but have not adequately sowed the means of future progress. Many of those manufacturers and others would, however, willingly give money towards such an object if they understood the value and the necessity of scientific research.

Whilst also many millions of pounds are annually expended in this country upon religious, philanthropic and other good objects, there is scarcely a scientific society or institution (with the exception of the Royal Society and the British Association) which expends even the small sum of five hundred pounds a year on pure experimental research in physics or

---

* In the year 1870, a gentleman of the name of Davis bequeathed £2,000 to the Royal Institution, London, to aid original scientific research.

chemistry. In the Royal Institution of Great Britain, the average annual expenses relating to experimental research, including salaries to assistants for research in the laboratory, from the year 1867 to 1871, did not amount to two hundred and fifty pounds. On the other hand, the "total net receipts" of the British and Foreign Bible Society alone, amount to about £213,000 a year. These circumstances strongly indicate extreme ignorance of the value and necessity of new scientific knowledge, and an equally strong desire to aid any good object which is understood. The money given to charitable and religious objects is largely a result of the unpaid labours of scientific investigators in the manner already described. The fact that verifiable truth is seriously neglected, whilst millions of pounds are annually devoted in this country to the support of dogmas and doctrines, proves that the English nation is even now in a very imperfectly civilized state.

Considering the multiplicity and variety of philanthropic institutions and bequests in this country, and the great effect original scientific research has in ameliorating the condition of mankind, and reducing the amount of human misery, it is surprising that no wealthy philanthropic individual has bequeathed funds for the endowment of an institution for pure research in physics or chemistry.* In

---

* As a notable exception to the above statement:—"Scientific research has now an Institute of its own in Birmingham, without

America, the Smithsonian Institution" was founded at Washington by benevolent and patriotic persons,* "for the increase and diffusion of knowledge among men," and one of the objects of that institution is "to enlarge the existing stock of knowledge by the addition of new truths," and a portion of its plan is "to stimulate men of talent to make original researches by offering suitable rewards for memoirs containing new truths," and "to appropriate annually a portion of the income for particular researches."

What is the reason that scientific research is not sufficiently encouraged in England? It is chiefly ignorance. There are very few good and important subjects, understood by the public, which are not in this country greatly assisted, nor many valuable public servants, whose labours are understood, who do not receive liberal payment and reward; and scientific research and discoverers therefore are neglected, not wilfully, nor because persons are unwilling to encourage good objects, but because scientific discovery and its great value to the nation are so little known. Scarcely a member of our legislature, or of our Universities, is fully acquainted with the national importance of scientific discovery,† and it

being indebted to the public funds. A fund has already been collected for carrying on the work. The building is called 'The Institute of Scientific Research.'" See *Nature*, January 7th, 1881, p. 366; the *Athenæum*, February 5th, 1881, p. 204; the *English Mechanic*, p. 537, February 11th, 1881.

* Professor Bache left 50,000 dollars, and Smithson bequeathed 541,000 dollars to this Institution.

† Respecting the Members of our Houses of Legislature, a former Postmaster-General remarked to me, that a dose of scientific research would be too much for them.

would probably be impossible to find a subject of such great magnitude so little understood. Comparatively few persons have clear ideas of the essential differences between scientific instruction and research.

Scientific research can only be successfully pursued by employing the highest motive—viz., a love of truth in preference to all things; and this is a condition which very few persons really understand, and a principle which a still smaller number practise. Men in this country are so accustomed to be actuated by the less noble motive of immediate self-interest or of some apparent practical result, that they cannot perceive that in scientific investigation the most valuable results can only be obtained by employing the highest motive. However necessary and effective the motive of immediate self-interest or of apparent practical result may be in ordinary affairs of life, it will not enable a man to make many discoveries, because it leads him away from those which are possible to search for others which may or may not be possible. The beginning of discoveries are often so very small, that it requires acute senses and observation in order to perceive them; and if the mind is preoccupied with a desire to discover some particular practical object, new phenomena are overlooked. In discovery, man must follow where Nature leads.

Another cause of want of encouragement of research, is the natural selfishness which exists, though in very different degrees, in all men. Many wealthy persons wish things to remain as they are. Some manufacturers would not aid research unless they could monopolize its advantages. Students also generally prefer those subjects which are best rewarded, and do not sufficiently consider their intrinsic value. The love of truth for truth's sake alone is very weak in most men, and but few men make the greatest good their chief object in life.

The extreme ignorance in this country of the value of scientific research, is also largely due to the narrowness of the "practical" character of the English mind ; men cannot perceive the deep-seated and universal sources of their wealth, and they prefer those occupations which yield the most obviously remunerative results. It is also partly due to scientific investigators themselves not having pleaded their own cause ; such men have been so absorbed in the more important occupation of discovery, that they have, probably more than any other class of persons, neglected to enforce the just claims of their own subject. It is, however, chiefly caused by the influence of misapplied wealth, operating through the old Universities and large public schools. The sons of the wealthy are most of them educated at those institutions, and according to evidence supplied by University authorities to Royal Commissioners, many persons send their sons to those places for other purposes than to acquire learning, and allow them too much money. The considerable wealth of these young men supplies them with attractions which decoy them from industrious study, and the wishes of the parents and students have been largely acquiesced in by the tutors and college authorities. At our old Universities also, physical and chemical knowledge is very much less rewarded than some other subjects, though latterly a considerable improvement has been made in this respect, but even now there is not a University in the kingdom in which a knowledge in scientific research is necessary in order to obtain the highest scientific honour.* In these various ways physical and chemical science has been kept very low in our chief seats of learning; and scientific research is greatly neglected by the governing authorities.

It is reasonable to suppose that Universities should be fountains of new theoretical scientific knowledge, as well as be the disseminators of it, and that they (especially the old ones with their rich endowments) would be certain to promote scientific research, as being especially a part of their functions; but such is not the case. Our old Universities have not established any professorships of original research; they make no payment for such labour, nor reimburse any expenditure incurred in such occupation, and afford but little facility for the prosecution of pure scientific inquiry. Further, they discourage scientific discovery by giving the greatest emoluments, and the

---

* The Victoria University has recently become a partial exception to this statement.

highest honours in science they have to bestow, to young men who have never made a single original research, or discovered a new fact in science. The money paid in the form of comparatively sinecure fellowships, or retiring pensions to young men in Oxford alone, "now amounts to about eighty or ninety thousand pounds a year." It may be objected that young men are not capable of doing original research, but as they do it in German Universities, they can also do it in England, if they are properly disciplined, and are not decoyed from industry by the possession or expectation of wealth. A man who has never made a scientific research is not the most worthy recipient of the highest scientific honours, and in Germany it would not be given to him; he is not properly disciplined in the detection of error or the discernment of truth in matters of science; he is deficient in accuracy of scientific judgment, and in the true spirit of scientific inquiry.

It is unnecessary to speak of what has been done during the last few years at our old Universities and great public schools, in the erection of laboratories, and in other ways for the promotion of science, because it has been for the purposes of instruction, and not of original research. No amount of ordinary instruction in science will remedy the evils caused by want of original inquiry, because such instruction does not produce new knowledge, but only disseminates that already possessed.

Many persons in this country think that all scientific men are investigators, and that a portion of the funds of scientific institutions generally are expended upon investigation, but such is rarely the case. Many also consider that those scientific men who are applying new knowledge are discovering new truths. And nearly all persons look upon inventors as the only really practical scientific men, and upon discoverers as unpractical enthusiasts who spend their lives in pursuit of vague theories. But whilst the inventor is a great and useful agent of civilization, there is one behind him who is greater than he, viz., the man who provides him with the new knowledge upon which all his inventions must be based.

The general aspect in which scientific research is viewed by many persons in this country, is that of a refined intellectual pursuit, which may be encouraged and honoured for the purpose of maintaining the tone of society. The question, however, is not whether this nation shall encourage research as a refined intellectual occupation, but whether it will contribute towards its own welfare by aiding scientific discovery.

Many persons also look upon scientific research as a hobby or as unpractical, and upon discoverers as mere accumulators of knowledge, but this is simply in consequence of their ignorance of the subject; if discoveries were commercial commodities, the practical character of research would be within their comprehension. A man who discovers knowledge for the use of invention is quite as practical a person as he who converts that knowledge into inventions fit for practical uses. The men who thus lead practical men must be practical themselves. Scientific discoverers may be considered the most practical men in existence, because their labours give rise to greater and more numerous practical results than those of any other persons. The discovery of a single substance, such as oil-of-vitriol, or washing-soda, has led to the formation of many valuable inventions, patented or otherwise, and to the establishment of thousands of manufactories. It is well known also that scientific discoverers are ardent lovers of truth, and are therefore very willing to communicate their knowledge for the good of mankind, and that manufacturers, men of business, and others, not unfrequently obtain from them and from their published researches, information of great value to themselves without even expecting to pay for it; forgetting that a scientific man may communicate in a passing remark, information which cost him years of labour to obtain.

Some persons also think that science is changeable and uncertain—that the discoveries of one generation are disproved by those of another, because they occasionally see scientific theories altered and superseded. But the real truth of the case is that the changes in the aspect of science which we continually witness do not often result from *alterations* in our

stock of positive knowledge, but from *additions* made
to it. Demonstrable truth is imperishable. It is true
that many theories have been invented and entertained for a while in the minds of scientific men,
and have then passed away, but we must remember
that these are only the scaffolding of science, and no
part of its real fabric. They consist of ideas which,
whilst they assist us in understanding science, and
in making discoveries, form no real part of our
positive knowledge.

Other persons seem to think that the laws of
matter are different in the laboratory from what they
are in the workshop; that the principles which
regulate a scientific experiment are different from
those which govern a large manufacturing process;
but this is a wrong idea. The laws of matter are
universal, substances have nearly the same properties in all places and in the hands of all men; water
boils at the same temperature whether in the retort
of a chemist, the saucepan of a kitchenmaid, or
the pan of a soap-boiler; iron wire is as readily
deprived of its rust in a chemist's acid bottle as in a
wire-drawer's pickling tub; a piece of phosphorus
will as readily ignite in the hands of a chemist as in
those of a match maker; a galvanic battery yields
the same quantity of electricity whether it be in the
hands of an experimentalist or in those of a working
electro-plater.

It is true that many things which have appeared
very promising in theory or in experiment, have

failed altogether in practice, but why is this? it is not that the principles of nature operated in the one case and did not operate in the other, but that we have imperfectly understood them, that from some unforeseen circumstances we have been unable to apply them; or that we have indolently abandoned them without sufficient or proper trial. In many cases we are unable to obtain the same conditions of success upon the large scale that we have upon the small one. In other cases a process fails because of its too great expense; many attempts have been made to supersede steam as a motive power by means of electro-magnetism, and engines driven by that force have been constructed of five or ten horse-power, but the cost of driving them has been found to be at least ten times the amount of that of the steam-engine of equal strength. And in other cases we fail because we attempt *at once* to carry out upon a large scale that which has only been the subject of limited experiment, instead of enlarging the process by small degrees, and adapting the apparatus, the materials and the treatment, to the size of the operation.

That also which appears very simple in the hands of an experimentalist, almost invariably becomes much more complex when carried into practice in a manufactory, simply because there is then a greater number of conditions to be fulfilled. Electro-plating a piece of steel with silver is to a chemist a very simple matter, because it is of no importance to him whether the silver adheres firmly, is of good colour, or is deposited at a certain cost; but with a *manufacturer* unless *all* these conditions are fulfilled, the process is a failure. These matters, however, belong to invention and not to original discovery.

We should not condemn theoretical science because we are not able, even with fair and persevering trial, to apply it to any useful purpose, but wait patiently until circumstances ripen for its application. Many inventions which are inapplicable in one state of knowledge become applicable by the progress of scientific research. The idea of an electric telegraph, attempted by Mr. Ronalds, in the year 1816, with the aid of frictional electricity, had to wait the development of the galvanic battery and the discovery of electro-magnetism before it could be successfully applied.

Many manufacturers seem to think that because some of their operations are completely routine, and have been handed down to them by their predecessors in nearly their present state, they are not at all indebted to science; but there is no manufacture, especially among metals, which has not in some degree been aided by scientific discovery.

In addition to the great benefits accruing from original research to all classes of society, our Governments have also derived immense advantages from the same source. The revenues have been greatly increased by the universal advantages conferred upon all kinds of industry and commerce by scientific knowledge. The additional taxes upon increased incomes from agriculture, arts, manufactures, mines; increased value of land and rents; investments in railway, telegraph, steam-ship and other companies, have been extremely great. From the sale of patents alone, a surplus sum of nearly six hundred thousand pounds has already accumulated. Our Governments are also indebted to original research for the use of percussion-powder, gun-cotton, improvements in cannon, projectiles, rifles, armour-plated ships, the ocean telegraph, field telegraph, the telephone, rapid postal communication, the speedy transport of troops and war-material, and a multitude of other advantages. The value of science to Governments in the prevention of war by means of more ready correspondence through telegraph is incalculable. Mr. Sumner, of America, at the period when the Atlantic telegraph was first employed, stated that the use of that telegraph averted a probable rupture between Great Britain and America. There was a period when we did not possess such evidence of the great value of science; but that time has now passed away, and our governing men have had abundant proof of the national importance of scientific discovery, and of the essential dependence of the welfare of this country upon scientific research.

Whilst vast sums of money are spent upon the applications of science in military and naval affairs, research itself is neglected; the superstructure is attended to, but the foundations are left to decay. A very small proportion of the money which is expended upon military affairs would, if devoted to research, save a great deal of expense in warfare:—

> " Were half the power, that fills the world with terror,—
> Were half the wealth, bestowed on camps and courts,
> Given to redeem the human mind from error,
> There were no need of arsenals nor forts."—LONGFELLOW.

Our Government has as yet made but little payment for the labour of pure research in experimental physics or chemistry; it has, however, given four thousand pounds a year for five years to be distributed by the Royal Society among scientific investigators, partly as personal payment. Income tax is deducted from these grants.

Want of recognition of the value of science has been so general in this country, that it is quite pleasing to quote a somewhat different case from the *Illustrated London News*, January 4th, 1873, viz., that of the late Archibald Smith, L.L.D., F.R.S. That gentleman was an investigator in pure mathematical science, and devoted the latter part of his life to the *application* of mathematics in the computation, reduction, and discussion of the deviation of the mariners' compass in wooden and in iron ships, and made practical deductions therefrom in the construction of those vessels. He published those practical applications of his scientific knowledge in the form of an Admiralty Manual, which was afterwards reprinted in various languages. Her Majesty's Government subsequently "requested his acceptance of a gift of two thousand pounds, not as a reward, but as a mark of appreciation of the value of his researches, and of the influence they were exercising on the maritime interests of England and the world at large." The kind of labour rewarded in this case was not scientific discovery, but the practical application of previously existing scientific knowledge.

The case of the late Dr. Stenhouse, F.R.S., is one of rather an opposite kind. That gentleman devoted his life throughout to pure investigations in organic chemistry, and published several of his researches in the Philosophical Transactions of the Royal Society.* His discoveries are very numerous, and although not much applied to practical uses by himself, the result of his researches on Lichens, and the yellow gum of Botany Bay, have been applied extensively by other persons in the manufacture of "French purple" and picric acid, and will doubtless continue to be applied to valuable uses. He held the Government appointment of Assayer to the Royal Mint, London, an office for several years unprofitable to him, but of increasing remunerative value, and which would have been subsequently worth £1,200 a year; but after the decease of his colleague, Dr. Miller, in 1870, that office, which was then worth to him about £600 a year, was abolished by the Chancellor of the Exchequer, and he lost the

---

* See "Royal Society Catalogue of Scientific Papers," vol. 5, pp. 719 and 890; and vol. 8, p. 1,010.

appointment, receiving, however, £500 as compensation. An application was therefore made to the Government, and a partial recompense to him was obtained, by Her Majesty granting him one hundred pounds a year "for eminence in chemical attainments, and on account of loss by suppression of office in the Mint." The only difference in these two instances, was, that in the second there was a very much greater amount of pure research and discovery, and a much smaller degree of applied knowledge.

These instances illustrate the statement, that however great an amount of valuable knowledge in pure science a man may discover and publish, or however freely he may provide others with the materials of invention and wealth, if he never invents anything, nor applies his knowledge to useful purposes, he is usually less rewarded even than an inventor. "The more intrinsically valuable the labour, and the greater the degree of profound original thought required to direct it, the less is it usually appreciated by the governing men of a nation." Absorbed in exciting questions relating to political emergencies, and national matters requiring immediate attention, even men of great administrative ability fail to appreciate the less direct though more fundamental sources of a nation's happiness and wealth. In harmony with these instances also, we find that it is not the pure sciences, but the concrete and applied ones, such as meteorology, geology, natural history, &c., in the Meteorological Department, the

Geological Survey, the British and South Kensington Museums, the Geological Museum, &c., and the National Gallery of Art, which have received the greatest degree of support from our Governments.

That discoverers are not treated by us as we treat other valuable members of the community is quite clear; either a physician, a judge, divine, lawyer, or railway superintendent of high ability, obtain from one to many thousand pounds a year, but a discoverer in pure physics or chemistry is, in scarcely any case, paid anything for his labour. That most eminent discoverer, Faraday, received for his scientific lectures at the Royal Institution of Great Britain, only £200 a year and apartments, during many years, and absolutely nothing for his great discoveries; and during the remainder of his life he only received a few hundred pounds per annum, including a pension of £300 pounds a year from Government. In contrast with this, the general manager of the Midland Railway has £4,000 a year. A General of our army receives £2,000, and a Field Marshal £4,000 a year (See "Whitaker's Almanack," 1873, pp. 121 and 138). A Head Master of either of the great public schools obtains from £3,000 a year upwards. An Archbishop of Canterbury receives £15,000 a year, besides a great amount of influence and power in the form of patronage to 183 livings, a palatial residence, and a seat in the House of Peers. A Bishop of London has £10,000, the patronage of 98 livings, and a seat in the House of Lords. I do not, however, mean to imply that these large emoluments are not deserved. Whilst also there are nearly 13,000 church benefices in England (See the "Clergy List," also "Whitaker's Almanack," 1873, pp. 153 and 155, and "Walford's County Families," 1872, pp. 173 and 610), there is scarcely a single appointment entirely devoted to scientific discovery, nor a single professor-ship in original research in science. I leave my readers to judge to what extent these instances illustrate the statement that discoverers are not treated by us as we treat other valuable members of the community. Partly in consequence of the fore-going neglect, the proportion of persons wholly devoted to scientific research in this country probably does not much exceed one in one million of the population.

It is scarcely credible that in a wealthy and civilized country, whilst the non-productive classes are protected in the enjoyment of titles and material wealth which in many cases they have not earned, the greatest scientific benefactors of the nation are constrained to live in straitened circumstances whilst working for the pecuniary and other advantages of those classes, and of manufacturers, capitalists, land-owners, and the nation in general. By these remarks it is not intended to imply that discoverers are inten-tionally neglected; but that the injustice they suffer is a disgrace to this country, and reflects discredit upon the governing classes, and especially upon those who reap the greatest advantage.

The men who are rewarded highly in this country are not always those who yield the greatest service to the nation, but frequently those who render the most immediate or most apparent benefit; to stop short at this cannot produce the greatest degree of success. The national services of a great discoverer are probably not equalled by those of any man. Who can estimate the value of the commercial, social, moral, political, and other great advantages to the world, of Oersted's discovery of the principle of electro-magnetism, which enabled the invention of the electric telegraph to be made? The men we reward the highest are not those who discover knowledge, but those who use or apply it; physicians, judges, bishops, lawyers, railway managers, military and naval officers, and head masters of schools, all of them gentlemen who render great services to the nation, by using, diffusing, and applying knowledge already possessed.

It requires less rare ability to apply knowledge to new purposes by means of invention, than to discover it; it is still less difficult to diffuse it by means of tuition and lectures, because the labours of a teacher consist largely of a repitition of other men's discoveries and inventions; and to use scientific knowledge in the ordinary business of every-day life, requires a still more common degree of ability.

A chief reason why ordinary business capacity is paid for whilst original research is not, is the fact that research is not considered a necessity; many

persons do not perceive its immense future value.
Men perform those duties first which they feel they
must : they are also willing to pay for the performance of those duties which press most urgently upon
them, and defer all other kinds of labour that they
consider will bear pos ponement. Most men act upon
this rule, until they acquire a habit of sacrificing the
future to the present, of neglecting more important
matters in order to attend to less, and of living too
much for money, without sufficient regard for the
more valuable condition, viz., individual and national
improvement.   These circumstances also largely
explain the fact that it requires more pressure to
induce individuals or governing bodies to aid original
research than to assist any other good object.   Other
chief reasons why persons in general cannot perceive
the great practical value of new scientific truth are,
because the perception of it requires a scientifically
trained mind.   The greatest truths are frequently
the least obvious, and are therefore valued the least.

It may be objected that research is not aided,
because it sometimes takes a long time to acquire a
practical shape and make it pay. We do not omit
to plant an acorn because it requires many years to
become an oak ; we do not neglect to rear a child
because he may not live to become a man ; but we
leave scientific discovery to take care of itself. The
intense desire which exists in this country for "quick
returns " has shewn itself in the much greater readi-
ness to aid technical education than to promote

permanent progress by means of original research. But the discoveries made in such a place as the Royal Institution of Great Britain have had a vastly greater beneficial effect upon civilization than that of any technical institution which has ever existed.

In a letter received by me from the Duke of Somerset, and which I have permission to publish, the true state of things in this country in relation to pure research is stated with remarkable accuracy and brevity :—

" The hindrances to scientific studies in this country are very many. The gentry are almost invariably educated by the clergy, and the clergy have seldom had time or opportunities for any scientific study. They usually take pupils or become tutors as soon as they have taken their degrees, and can only teach the Latin and Greek which they have themselves learned. The commercial classes value what they call practical science; this means some application of science for the purpose of making money. Competitive examinations may promote a superficial acquaintance with the elements of science, but are unfavourable to the development of scientific culture. The scientific associations tend to degrade science by exhibiting scientific men as candidates for applause from assemblies which seek amusement and startling results from lectures and experiments. The advancement of science, is therefore, left to comparatively few men, who are unregarded and unrewarded."

To remedy this state of things we require a general encouragement of pure scientific inquiry by the State and Universities. It is thought by some persons who have given special attention to the subject, that the State ought to encourage such research and science in general, by appointing a Minister of Science possessing scientific knowledge and good administrative ability ; a Scientific Council to advise our Governments in all important matters relating to science ; and by establishing State laboratories for pure scientific inquiry, with discoverers of repute in them wholly engaged in research in their respective subjects.

There are also many new experiments, investigations, and explorations, which neither private individuals, nor even corporate bodies, such as the Royal Society, the British Association, Geographical Society, can effectually make, and which only a Government can carry out, such as Arctic expeditions, trigonometrical surveys, deep sea dredging operations, magnetic observations, determinations of longitude, meteorological and astronomical observations, researches on tides, observations of earthquakes, determinations of the height of mountains and the density of the crust of the earth, experiments on the best form of ships, geographical explorations, and many others.

It is clear from the enormous advantages which this nation has already derived from scientific discovery in physics and chemistry, pursued with only the aid of the very limited means of private persons, that had research in those subjects been sufficiently supported, the manufactures, arts, commerce, wealth, and civilization of this country would have been much greater than they are; emigration also of the industrious classes, disease, pauperism, crime, the evil effects of famine, etc., would have been much less. The amount of knowledge and riches obtainable by means of research and invention is practically unlimited, and it is astonishing that this immense source of industry and wealth in a nation should have been so neglected by our Governments. The practical value of new scientific knowledge is vastly greater than that of all our goldfields or even of our coal supply, because it would not only enable us to obtain from coal several times the amount of available heat and mechanical power we now secure, but also to apply to our wants the numerous other materials composing the crust of our globe and the contents of our oceans; also all terrestial forces, the internal heat, the tidal energy and atmospheric currents, and the immense amount of power this Earth is continually receiving from the Sun. Whilst at present vast amounts of materials and energy remain unutilized, nearly all those terrestrial substances and forces might probably be rendered of service to us if we possessed sufficient knowledge.

That scientific research is a far greater source of wealth and wellbeing than our stores of coal is easily proved. At present we obtain in our best steam-engines only about one-seventh (or less) of the mechanical power producible by the combustion of the coal, the remainder being lost in various ways. And this occurs simply because we have not yet discovered a method of wholly converting heat into mechanical power. In some other instances we are able to convert one force wholly into another without loss, as for example : the chemical action of a voltaic cell into electricity; and by means of research we shall probably be enabled to effect a similar complete conversion of other powers into each other. The effect of converting heat wholly into mechanical power would be equal to increasing our stock of coals for that purpose to seven times its present amount. This instance is only one of the many thousand possible ways in which research may yet prove of value to mankind.

It is true that a very large amount of original research in physics and chemistry has been done in this country; the contents of our scientific journals and of the publications of our various Learned Societies prove this. It is also true that the English nation has been pre-eminently active in applying scientific knowledge to practical uses by means of inventions, and has been generally the first in carrying out inventions on a large scale. We have been either the first, or nearly so, in developing steam-engines, railways, locomotives, rapid trains, gas works, flour mills, blast-furnaces, cotton machinery, cheap postage, light-houses, electro-plating, lucifer-matches electric-telegraphs, submarine electric cables, great engineering establishments, iron ship-building, and many other important enterprises. Three out of four of all the great ocean steamers, and three-fourths of all the locomotives of the world were constructed in this country.* By means of our enterprise and capital also, the first railways, telegraphs, gas works, cotton mills, modern water works, suspension bridges, water wheels, harbours, light-houses, &c., &c., in nearly all parts of the world were constructed ; and foreign nations have been inducted into the practical methods of working our great manufacturing and technical applications of science.

By means of English enterprise and skill the cities of Aix-la-Chapelle, Altona, Amsterdam, Antwerp, Berlin, Bordeaux, Brussels, Cologne, Frankfort-on-Maine, Ghent, Haarlem, Hanover, Lille, Rotterdam, Stolberg, Toulouse, Vienna, and others were lighted with gas. We formed Water Companies or Waterworks in Amsterdam, Berlin, and other cities, and drained Naples. We utilized the falls of the Rhone at Bellegarde, and thus obtained 10,000 horse-power for the use of the French manufacturers. We also sent the first steam-boat to Coblentz in 1817, and the first to America. We laid the first Atlantic cables. And as a general truth, we have been foremost in invention, application and enterprise.

---

* See *Nature*, April 24th and May 1st, 1873, pp. 485 and 13 ; also *Work and Wages*, by Brassey, pp. 170 and 178.

Recent International Exhibitions however, and the migration of various branches of our trade to the Continent and America, have shown that the degree of our relative superiority in manufacturing skill is diminishing. Other nations, especially the German and American, perceiving the dependence of invention upon research, and the enormous pecuniary and other advantages gained by us, by the application of scientific knowledge to manufacturing and other purposes, have within the last few years aroused themselves, and are now pursuing pure science much more energetically than ourselves. A few years ago the relative number of original researches made per annum in England, France, and Germany were in the proportion 127, 245, and 777. Many of those made in Germany were valuable ones, and were made by Students in order to obtain a degree. Other nations are rapidly gaining upon us in the application of science to industrial purposes, and have even surpassed us in the extent of some of their manufacturing and technical operations. Many persons who have visited Europe and America at intervals during the last twenty years have testified to this.

The Vielle Montagne Zinc Company in Belgium employ 6,500 workmen, and produce annually 32,000 tons of zinc. The John Cockerill Company, engine-builders, Seraing, near Liege, employ nearly 8,000 men. Krupp, the great engineer at Essen, near Dusseldorf, employs about 10,000 workmen; his works at Essen alone cover 450 acres, and 1,000 tons of coal are consumed in them daily. The Anzin Company (Valenciennes) "is the largest coal company in the world, producing no less than 1,200,000 tons per annum, and employs 8,000 hands." The Chatillon and Commentry Iron and Coal Company (France), produce annually from 300,000 to 350,000 tons of coal and coke, nearly 70,000 tons of iron and steel, and employ nearly 9,000 workmen. At the Creuzot Ironworks (France), "the mineral concessions cover an area of nearly six square miles, the coal-fields nearly twenty-five square miles, the building 296 acres. There are nearly forty-five miles of railway between various parts of the works, upon which are generally running sixteen locomotives. The galleries in the mines are more than twenty miles long." 10,000 persons are employed in the works and the annual amount of wages paid equals £400,000.*

Our practice with regard to original science has been very different from the plan carried out in Germany. Within the last few years great laboratories have been erected in Berlin, Leipzig, Aix la Chapelle, Bonn, Carlsruhe, Stuttgardt, and other places, at the expense of the State, and special provision has been made in them for original scientific research. A glance at the frequently published list of scientific investigations made in different countries will shew us that the Germans have been making a far greater number of discoveries in science than ourselves.

Sir R. B. C. Brodie, Professor of Chemistry at

*Note.—See "Work and Wages," by Brassey, p.p. 15-131 and 132 ; also the "Laboratory," vol. 1, p.p. 313-316-378 and 380.

Oxford, speaking of his experience when a student at Geissen, in Germany, states : " I say that the enthusiasm and earnestness of the young men in the laboratory was quite unparalleled in my experience at Oxford. The dilettante sort of way in which things go on there is very inferior indeed to the way the German students study. At Heidelberg, I have been told, there are about eighty professors, and amongst those professors are some of the most eminent men in Europe, so that they have a staff quite unsurpassed."

The industry of the Germans in scientific research is quite remarkable, they are availing themselves of the great fountain of knowledge to a much greater extent than ourselves, and are already beginning to reap the reward. Within the last few years they have succeeded, by means of researches, in making alizarine, the colouring principle of madder. "England produces immense quantities of benzene, the greatest part of which goes to Germany, there to be converted into aniline dyes, a considerable quantity of which goes back to England. No other country is so far advanced in the manufacture of the coal-tar colours as Germany. The quantity of alizarine manufactured by the German makers far surpasses the English production." (See "Alizarine, Natural and Artificial," by F. Versmann, New York, 1873). Statements of this kind are frequently published, and made by our manufacturers and others, of the departure of branch after branch of our manufactures to the Continent, and of continually increasing importation of foreign-made articles.

Some persons, having become aware of the cosmo-politan nature of scientific research, have suggested that it is a matter of no importance to us as a nation whether we make researches or not, as foreigners would make them, and we could apply them. But no honourable man would, after reflection, seriously main-tain such a proposition, because it implies a willingness to obtain from the labours of other persons, advantages without paying for them. It is partly this absence of a desire to pay for the labour of investigation, which is now damaging the manufacturing and commercial prosperity of this country. It is also certain that how-ever much we may have hitherto succeeded commer-cially, without making payment for research, we should have succeeded much better had we properly assisted investigators in pure science. Our success has hither-to been obtained, not in consequence, but in spite of the disadvantageous circumstances under which dis-coverers have laboured.

The commercial argument in favour of encouraging research, although the most effective with the great mass of persons, and therefore much dwelt upon in this chapter, is however quite a secondary one; the en-couragement of truth for the sake of its own intrinsic worth, in preference to the material or extrinsic value of its results, should be the foundation of all aid to discovery. Justice, also, ought to come before all minor considerations, and no upright man would wish for a moment that anyone, and much less the greatest scientific intellects in the country, should work for his benefit without being remunerated.

- It has been objected that Continental nations, the Germans in particular, have pirated our patents, infringed our designs, imitated our labels, used our names, and taken our improvements wholesale, and this may be true. But we still have had by far the largest portion of the reward of our greater energy and inventive skill ; we have had the great advantage of being first in the markets of the world ; and that advantage can only be retained by our being the first in the pursuit of original research, as we have so long been in the application of science to industrial arts, and not by purchasing foreign inventions, nor by accepting gifts of unrecompensed researches.

Nations as well as individuals are apt to push to an extreme the means by which they have succeeded in gaining either riches or power. We have devoted ourselves relatively too much to the pursuit of money and too little to the pursuit of knowledge. The desire for wealth is in this country so great, that probably nothing but a loss of that wealth will ever make us properly encourage the pursuit of new knowledge.

Whilst research is being neglected, manufacturers and others in all directions are asking for improvements in their machines and processes ; employers of steam engines want to obtain more power from the coals; makers of washing soda wish to recover their lost sulphur ; copper smelters, want to utilize the copper smoke ; glass makers wish to prevent bad colour in their glass ; iron puddlers want to economise heat ; gas companies are desirous of diminishing the leakage of gas; iron smelters wish to avoid the evil effects of impurities in the iron; manufacturers in general want to utilise their waste products and prevent their polluting our streams and atmosphere; and so on without end. And inventors are continually trying to supply these demands, by exercising their skill in every possible way, with the aid of scientific information contained in books; but after putting manufacturers and themselves to great expense, they very frequently fail, not always through want of inventive skill, but often through want of *new* knowledge attainable only by means of pure research. Judging from the vast amount of inventive skill already expended upon the steam engine, and the small proportion of available mechanical power yet obtained from the coals consumed in it, it is highly probable that a machine for completely converting heat into mechanical force cannot be invented until more scientific knowledge is discovered.

It must not be supposed from these remarks, that discoveries which will enable a man to make any particular invention, can be produced to order; that is only true to a very limited extent. Men are beggars of nature, and must not expect to be permitted to choose her gifts, or dictate what secrets shall be disclosed. We may however be certain that if we acquire a very much greater supply of new scientific knowledge, we shall then be able to perfect many good inventions, though not always of the kind we wish, or in the way we expect. The great sewage question may perhaps be solved in quite an unexpected way, possibly by the discovery of some substance capable of precipitating ammonia and organic matter from their solutions.

Nearly all our manufacturing processes are full of imperfections; thus the loss of gas by a single large provincial gas company, after that substance has left the works, amounts to nearly one hundred and fifty millions of cubic feet per annum, and to a value of about £18,000; and the soil of all our large cities and towns is permeated and rendered fœtid by coal gas. And it has been stated by an eminent authority in such matters that we might save 500,000 tons of coal a year by economizing the waste heat of furnaces, by purifying the coal, coking it, etc. In a single chemical manufactory, out of about two thousand tons of hydro-chloric acid used per annum, about eight hundred tons have been allowed to flow away as a polluting substance, because it was not possible to utilise it. The loss of material from a single large glass works equals fourteen hundred tons per annum, and a value of £8,000. Similar grave defects might be pointed out in nearly all our large manufactures, by those acquainted with the subject.

Inventions are wanted for quickening the process of vinegar making, and diminishing the percentage of loss of the acid. For bleaching discoloured fats. For quickening the process of converting cast iron into malleable iron. To easily separate nitrogen from the oxygen of the atmosphere. To economically convert the nitrogen of the air into valuable products, such as nitric acid and ammonia. To find uses for the im- mense quantities of minerals which abound all over the earth ; to utilise wolfram and find applications for tungstic acid; to apply titanic acid to great industrial purposes ; to produce aluminium on the large scale, as we now produce iron. To tan leather more quickly, and without detriment to its quality. To prevent the rusting of iron. To more perfectly prevent smoke. To collect and use the sulphuric acid of the salt cake consumed in the glass manufacture. To make window glass by means of common salt. To deodorise offen- sive substances. To find larger uses for phosphorus, sodium, magnesium, and common salt. To remove phosphorus and sulphur from iron ores, and sulphur from coal and coke. To obtain a good white alloy as a cheaper substitute for German silver. To convert white phosphorus into the red variety by a less dan- gerous process than the present one. To prevent the putrefaction of "peltries" in glue making. To obtain better and cheaper materials for colouring glass. To more perfectly prevent animal food from change. To obviate or prevent explosions in mines. To perfectly purify ordinary red lead for making flint glass. A cheaper process for converting common salt into washing soda ; and so on without end.

We also very badly require a method of recording our thoughts in readable forms upon paper, without the slow and laborious process of writing. An incal- culable amount of brains and of intellect, especially of the greatest thinkers, would be saved by such a discovery. The curative arts also are permeated with empiricism, and thousands of lives of persons of all classes of society, are annually lost in this country through want of a more perfect scientific basis of medicine, attainable only by means of experiment and observation.

In this country, such great practical results have been obtained by means of invention, that many persons suppose a sufficiency of inventive skill will enable us to effect every possible scientific object, and are surprised that no one can invent a plan of utilising the entire heat of coals, or a mode of overcoming the sewage difficulty, or prevent the great leakage of coal gas, or arrest epidemics, or produce a steam engine which shall work without waste of power. The progress of invention however depends upon that of discovery, and these various inventions, etc., wanted by manufacturers and others probably cannot be perfected until suitable *new* knowledge is found. Every new invention has its own appropriate discoveries, by means of which alone it can be perfected ; it was not possible to perfect the idea of an electric telegraph before the discoveries of Volta and Oersted were made According to scientific laws, out of everything proceeds everything, and out of nothing, nothing can come, even ideas are not created. An unlimited number of inventions cannot be made by means of a limited amount of scientific knowledge ; and our present stock of such information applicable to inven-

tion, is very insufficient.   One great reason why only
a small portion of patents are of practical value ; and
so many useless ones are taken out is, that in consequence of our so-called "practical" spirit, we over-
estimate the power of invention and under-value the
discovery of new abstract truths ; because also invention has done so much, we think it will continue to do
so, but the latter depends upon a continued supply of
discoveries.

Nearly every manufacturer is aware by painful experience of the great and almost incessant variation
that occurs in the quality and properties of the materials used in his trade, and the frequent risk of failure
of his process.   In the manufacture of iron, for
example, the presence of much phosphorus, sulphur,
or silicon in the ore is liable to be very detrimental to
the quality of the iron produced from it; in the manufacture of glass, the least quantity of iron in the
materials will seriously injure the colour of the product;
in the selection of copper for telegraph wire, if it contains the least trace of arsenic, the wire will not conduct
the electricity properly.   The difficulties experienced
in procuring suitable materials for a manufacturing
process are in some cases very great ; and when they
are procured, additional difficulties arise from the inability of the manufacturer or his manager to analyse
them.

Every manufacturer is also aware that the difficulties
encountered in manufactures are not limited to the
substances employed, but extend to all the different

processes and stages of processes through which these substances have to pass, and to all the forces, tools, machinery, and appliances employed in those processes; in the manufacture of glass, for example, the greatest care has to be exercised in the making and gradual heating of the pots in which the glass is melted, the proportions of the materials, the construction of the furnaces, the management of the heat, and a whole host of minor conditions too numerous to mention, all of which must be attended to with the greatest care. In the manufacture of iron and steel, the smelting of copper, the refining of nickel, the preparation and baking of porcelain, and in many other trades, innumerable difficulties, all having their origin in the properties of matter and forces, continually beset the manufacturers. In some cases difficulties occur which perplex both the workman and the scientific man called in to his aid, and so far from an unscientific workman being able to overcome them, even with the aid of the scientific man, he is unable to do so.

The hidden difficulties which beset a manufacturer are not unfrequently so inscrutable that the present state of knowledge in science fails to explain them. Who can tell why it is that wire-work of brass or German silver becomes gradually brittle by lapse of time? Or why varnish made in the open country has different properties from that made in a town? Or why silk dyed in Lyons should possess a finer colour than the same silk dyed by the same process in Coventry?

With our present extremely imperfect knowledge of Physical and Chemical science, we can perhaps hardly form an idea of the amount of knowledge yet to be discovered respecting the phenomena which manufactures present.

One of the inevitable results of these difficulties in manufacturing processes and of deficiency of knowledge, is the production of a large amount of goods of an inferior quality; and useless goods, technically called "wasters," the cost of which has to be laid upon the saleable ones, and thus the price of the latter is enhanced to the consumer. For instance, flint glass discoloured by iron has sometimes to be sold at a loss for making common enamel; waste window glass has to be sold as "rockery" for ornamenting gardens, and defective articles of glass or metal have to be re-melted.

In consequence of this want of new knowledge, manufacturers continue to suffer losses which might be avoided; high prices of useful articles are maintained; defects in their quality are not improved; preventable accidents still continue to happen; the health of workmen continues to suffer; many means of curing diseases remain unknown; medical practice remains full of empiricism, &c., &c.

The great sewage question is apparently in this predicament; we are probably trying to solve it without first discovering the requisite knowledge; inventors, engineers, and consulting chemists have racked their brains, and have not been able to devise a satisfactory remedy, and meanwhile the health of the entire population of this country is suffering. If we so neglect the fundamental means of ameliorating our condition we deserve to suffer. One would suppose that cholera, contagious diseases, colliery accidents, pollution of air and water, enormous waste of heat from fires, and a multitude of other evils which depend upon physical and chemical conditions, are of but little importance, that we should so neglect one of the most effectual means of preventing them ; and it is perfectly clear that by neglecting to aid research, those who gain so much money and advantage from original science, and render no return, are unwittingly sacrificing national interests upon a large scale to personal benefit.

The practice of some manufacturers using and deriving great profit from new knowledge evolved by research, without recompensing the discoverers, sometimes causes injury to the public welfare by preventing the publication of discoveries which have an immediate practical application. Experience of this kind has constrained me to postpone the publication of a method I have found of readily and quickly converting lumps of white phosphorus into the red variety in a state of powder without protracted heat or grinding.

" What will be the next chapter of British enterprise and invention, and who and where the men to perform the chief part in it ? As to the work to be done, there can be no doubt or mystery, for not a day passes without loud complaints, indignant remonstrances, fatal oversights, sad mis-calculations, terrible short-comings, social or material evils to be remedied if possible, whole masses of people, indeed whole classes to be succoured and lifted out of the slough, and enormous difficulties placed by nature in our way evidently that we may exercise our wit and our virtues in the attempt to overcome them. Here, from all these Isles, there arises a despairing cry from agriculture, as if it had really reached the end of its tether, and had found itself landed in utter helplessness and insolvency—a bad speculation altogether. Here are countless problems, and at the same time countless discoveries, which if they lead to nothing else, prove the inexhaustible nature of our dominion over the elements. Then, for the sea, with its terrible average of wreck and total loss running on without intermission and with but rare abatement, who shall say there is here no work for the discoverer and inventor who will give his heart and soul and mind to it ? "

It is indeed high time, that by means of discoveries which will enable us to predict with certainty the nature of coming seasons, we shall be better enabled to cope with adversities in agriculture ; also, that the numerous wrecks, and the thousands of lives lost with them every year on our coasts, should be diminished. But these desirable results cannot be effected by invention based upon insufficient knowledge ; invention must be preceded by general as well as special research, because the former often discloses important truths which we cannot predict. Our present electric lights in light-houses and on large ocean steamers, had their origin, not in direct inventions or special researches for the purpose, but in abstract researches on apparently remote subjects.

It is nothing less than a national crime that proper provision has not yet been made for investigating scientifically the causes of famine and pestilence, also physiology and pathology, and the discovery of the laws which regulate diseases and epidemics. What can be more painful to behold than a mother and father deprived of a whole family of five or six children in rapid succession by scarlatina or other contagious disease, and both the parents and medical men utterly unable to save them ; and this is a common occurrence. Persons who are ignorant of science look with an abject feeling of helplessness upon great national calamities, and even upon private afflictions, such as a local epidemic, as if there was absolutely no remedy, whilst scientific men believe that by extension of knowledge, such evils might be largely avoided or prevented.

Many persons however, actuated by the very kindest of motives, but insufficiently acquainted with the necessity, conditions, results, and advantages of experiments, unwittingly obstruct the discovery of new knowledge in physiology and pathology, by attempting to prevent experiments being made upon animals.

We should not strain at a gnat and swallow a camel. Nearly every step in life involves a choice between two alternatives, and this is the case with experiments upon living creatures, either such experiments must be made, or the wholesale slaughter of men and other animals, by pestilences, epidemics, small-pox, foot and mouth disease, &c., must continue. Many of the properties of living bodies, like those of dead ones, can only be ascertained by means of experiments, no other course is possible ; and the knowledge so obtained enables us not only to prolong the lives but also to alleviate the sufferings of all kinds of living creatures. Nearly all our medical and surgical knowledge has been obtained by observation and study, either of the results of experiments made by ourselves, or by the course of nature for us ; and the former is often attended by immeasurably less pain and expence than the latter. No one who has ever made in a proper manner new experiments, would venture to assert that valuable knowledge is not gained by them ; and this statement is as correct of experiments in physiology as in all the other sciences.

The total amount of pain inflicted upon animals by vivisection experiments in this country is infinitesimally small—because, firstly, the proportion of experimentalists in so-called "vivisection," does not amount to one person in one million of our inhabitants :—secondly, students cannot be induced to enter upon scientific research in physiology, because such labour is unrewarded, either by enabling them to obtain certificates, degrees, or money. Whatever pain also, is inflicted in such experiments, is by men of the highest eminence in physiology, and therefore by the most competent persons.

Experimental research is an occupation requiring an exceptional kind of ability and experience ; and persons who have never made experiments, nor studied their relation to human welfare, are largely incompetent to determine when and how they should be made, the real effects of them, or the value of the knowledge they afford. To persons inexperienced in scientific research, many experiments appear useless, which have great practical value, either immediately or at a later period. Our greatest curse is ignorance ; and knowledge, by enabling us to avoid the fatal effects of pestilences, and epidemics, is as necessary as food to mankind. The " Anti-vivisection " movement however is but one of the phases of the ever-existing conflict between the advancing and retarding sections of mankind.

Greater sympathy with suffering accompanies greater civilization. The increased humanity of the present age over that of previous ones, is largely due to the discovery and extension of new scientific knowledge. Science, by showing more clearly to man his true position in nature and in relation to his fellow-men and other animals, has rendered more evident the concrete fact, that the happiness of each depends upon the happiness of all, and the happiness and welfare of all upon that of each individual. It has also operated in a more apparent, though less important way, by inculcating better systems of hygeine, improved sanitary arrangements, &c., &c. It is not to the zeal of "anti-vivisectionists," but to the well-directed labours of experimental medical men, that mankind are indebted for the discovery and invention of nearly every known method of preventing and alleviating animal suffering and of prolonging human life. This statement is true of vaccination, the use of chloroform in general surgery, dentistry, and mid-wifery, of carbolic acid spray in surgical operations ; the abolition of the practice of searing amputated limbs with a red-hot iron ; and many other improve-ments. Ferrier's comparatively recent vivisection experiments have already enabled medical men to treat more successfully those formidable diseases, epilepsy and abcess of the brain.

What this nation badly requires, is not less experi-mental research, but more. When famines result from insufficiency of Solar heat, instead of investiga-ting the conditions of the Sun's surface to enable us to predict their occurrence and provide accordingly, we allow them to come upon us in our unprepared state and produce their fearful effects. When con-tagious disease overtakes us, what do we do? Instead of previously employing and paying scientific inves-tigators to make experiments in physiological and chemical science, to enable us to combat it success-fully, we vainly attempt to apply our present stock of chemical and physiological knowledge to ward off the difficulty. When high price of fuel intervenes, instead of previously giving discoverers the means of finding new principles relating to heat, and to chemical, and electrical action, we ineffectually endeavour by means of invention, to economise fuel.   These are the pottering, short-sighted, and ignorant ways in which " the great English nation" temporises with great evils, and permits national welfare to be sacrificed to private gain, instead of employing for the discovery of new knowledge some of that superfluous wealth which in many instances is a curse to its possessors.

CHAPTER II.

The Scientific Basis of Mental and Moral Progress.

It is not highly necessary after what has been already said in these pages, to adduce much evidence to show that scientific discoveries, either directly or through the medium of the inventions based upon them, have been a great cause of mental and moral progress. As however there are many persons who do not perceive the dependence of such progress, and especially of moral advance, upon science, a few of the chief relations of those subjects to each other may be pointed out.

The dependence of mental progress upon science may be rendered manifest in several ways :—1st. By showing that new scientific knowledge is continually extending and modifying our views of existing things. 2nd. That inventions based upon scientific discoveries have aided and extended our mental powers :—3rd. That mental phenomena may be made the subject of experiment, observation, analysis, and inference :—

*NOTE. —The whole of this chapter, especially the Moral Section, is capable of great amplification and much more copious illustration.

4th. That the criteria of truth, and the mental powers and processes employed for discovering and detecting truth, are the same in mental as in physical science, and, 5th. That mental action is subject to the great principles and laws of science. And moral progress may be proved to have a scientific basis:—1st. By shewing that moral actions are a class of mental actions, and therefore subject to the same fundamental laws and influences:—2nd. That the discovery of new scientific knowledge, and the use of inventions based upon it, often conduce to morality:—3rd. That moral phenomena may be made the subject of experiment, observation, analysis, and inference:—4th. That the criteria of truth, and the mental faculties and processes employed, in discovering truth, are the same in moral as in physical science:—5th. That the fundamental rules of morality are subject to the great principles of science:—6th. That moral improvement follows in the wake of scientific advance:—and 7th. By showing the moral influence of experimental research in imparting "the scientific spirit;" promoting a love of truth; dispelling ignorance and superstition; detecting error; imparting certainty and accuracy to our knowledge; inculcating obedience to law; producing uniformity of belief; aiding economy and cleanliness, promoting humanity, &c., &c. Each of these will be treated with extreme brevity,

MENTAL PROGRESS.

The chief object of this chapter is only to shew

that mental action is largely consistent with the great principles of science ; not that in our present state of knowledge, mental phenomena can be entirely explained by them, or that mental actions involve nothing more than physical and chemical processes.

That mental progress is advanced by scientific discovery is a common circumstance. Our ideas of facts, our knowledge of general principles, our views of man, of nature, and of the Universe ; and even our modes of thought, have been gradually and profoundly changed by the new knowledge acquired by means of scientific research. This truth is capable of being most extensively illustrated by a multitude of facts in the whole of the sciences, and in the arts, manufactures, and other subjects dependent upon science. For example, in astronomy, great changes, produced by the results of scientific discovery have taken place in our ideas respecting the magnitude of Space and of the Heavenly bodies, the constitution, form, and motion of the Earth, the functions of the Sun and Moon, the distances of the Sun and fixed Stars, the nature of eclipses and comets ; and a great many other matters. In terrestrial physics, the mental advances have been equally great in our ideas respecting the causes of tides and of winds, the pressure of the atmosphere, the existence and course of the Gulf Stream, the physical conditions of the Equator and Poles, the conditions upon which day and night, summer and winter depend, the depth of the ocean, the height of the atmosphere, the cause of rainbows, of rain, hail, snow, mist and dew, of thunder and light-ning, the composition of air, water, mineral, and organic substances, and other most numerous and varied phenomena.   In the subjects of heat, light, electricity, magnetism, chemistry, vegetable and animal physiology, psychology and morality, and the more concrete subjects depending upon them, such as politics, trade, commerce, government, &c., our ideas have equally advanced, in consequence of scientific research ; and to fully describe the mental progress resulting from discovery in nearly all branches of human knowledge would require a series of books to be written on the History of all the Sciences.

Other causes also, which I need hardly mention, besides scientific discovery, have of course contributed to the mental progress of mankind.   We arrive at true ideas, not only by the more certain and syste-matic process employed in scientific research, but largely also by the uncertain method of trusting to instinct and habit, by adopting dogmatic opinions, and by the semi-scientific plan of following empirical rules.

Dogma and empiricism, in nearly all subjects, has rendered immense service to mankind.   Contempora-neously with the progress produced by new knowledge, the mental condition of man has been maintained and prevented from receding, by the combined influence of hereditary mental proclivity, acquired habit, pro-mulgation of dogmatic opinions and empirical rules, and by previously known verified truth.   Religious belief has thus been the forerunner of Science. Dogma and empiricism are indispensable agents of civilization ; they cannot be dispensed with by the great mass of mankind, who have not the time at command, nor possess the other means, necessary for acquiring verified knowledge. They afford rough and ready guides and useful " rule of thumb " methods, though less certain and less accurate than those afforded by verified and definite science.

That various inventions, based upon scientific discoveries, have greatly aided and extended our mental powers is quite certain. The discovery of the properties of a mixture of solution of nutgalls and green vitriol, has, through the invention of ink, exercised an immense influence in promoting the mental developement of mankind ; and the discovery of the properties of esparto grass and other materials for making paper has contributed to this result. Every discovery also resulting in inventions which facilitated the transmission of intelligence has had a similar effect. Amongst these are magnetism, which, in the mariner's compass greatly assisted navigation and the conveyance of letters by sea ; and the steam engine which facilitated the transmission of letters by land and by water ; the electric telegraph, the telephone, and other contrivances for transmitting ideas, have also greatly promoted mental advance. The steam engine, by largely abolishing physical drudgery, gave time for study and mental and moral improvement. It has been said that " it is impossible to lay down a

railway without creating an improved intellectual
influence. It is probable that Watt and Stephenson
will eventually modify the opinions of mankind, almost
as profoundly as Luther and Voltaire." Photography
has exercised an immense intellectual influence of an
improved kind, by making common to all mankind,
views of the beautiful scenery of all parts of our globe,
and portraits of individuals of all nations and of all
classes of society. Processes of printing from electro-
type plates, pictures and letter-press, upon the paper
wrappers used by grocers and other tradesmen, have
also carried into the homes of millions of poor persons
truthful ideas and an improved intellectual influence.
The invention of steel pens, of which a thousand
millions are made yearly in Birmingham alone, must
also have considerably aided intellectual progress.
The various calculating machines used by merchants,
the copying presses, papyrographs, and the numerous
inventions for copying and multiplying letters and
circulars and for domestic printing, have saved intellectual toil, and promoted the diffusion of intelligence.
These are only a few of the numerous ways in which
inventions based upon scientific discoveries, have
resulted in mental progress.

Less perhaps has been done in the way of actual
definite scientific experiments upon mental actions
and processes than in almost any other department of
science, and this is partly accounted for by the fact
that the other sciences require to be largely advanced
before we can use them to examine mental action,

and partly also because (as occasionally happens) the latter has been a neglected subject of research. During the past few years however, various experimental investigations have been made, especially by Donders in Holland, and Mosso in Turin, for the purpose of elucidating the physical conditions of mental action ; and it has been found that instead of an act of thought being instantaneous, as was formerly believed, it requires a variable time.* Numerous desultory experiments made upon dreamers, and with drugs, alcohol, &c., upon persons in the waking state, also prove that mental phenomena are amenable to scientific research. F. Galton has even proposed experiments and methods for measuring the mental faculties of different persons.† The effects of exciting different parts of the brain of animals by means of electric currents, and the localization of the functions of the brain effected by the experiments of Ferrier, Hirtzig and others, also tend to throw further light upon mental phenomena. The fact alone that mental actions and conditions may be made the subject of experiment, and consequently of observation, comparison, analysis and inference, proves that they may be rendered sources of new facts and principles, and are therefore within the domain of science. As the dependence of mental phenomena upon physical conditions has been clearly demonstrated, an extensive reduction of them to scientific laws is only a question of time and labour.

* Note. — See also p. 95.   † Note.— Athenæum, Aug. 3, 1877. p. 242.

The principles of nature and the modes of mental action are the same for all men.   It necessarily follows from the essential nature of truth and the invariability of the chief methods of detecting it, that the criteria of truth in mental science, and the mental powers and processes by which truth is arrived at and detected in that science, are essentially the same as in the physiological, chemical, and physical ones.   In each of these subjects, we first, either with or without the aid of experiment, make observations, record facts, compare them, and draw conclusions from our comparisons ; we also group the facts, and the conclusions, in every possible way, and then draw other conclusions ; we also analyse, combine, and permutate the various truths arrived at, and cross examine the evidence in every possible manner in order to extract from it the greatest amount of new knowledge. And in each case we employ as the criteria of truth, the test of consistency with the whole of the evidence bearing upon the case, and especially with the great principles of science.   We determine what is true, chiefly by comparison with those principles, because they are the most firmly established true ones and the most universal.   There is no royal road to truth, and no special mental faculty for detecting it in any subject ; and it is in consequence of our mental faculties being so very finite that we have no easier way of arriving at truth.

No dogmatic teaching can ever, except by accident, fully explain to man the true nature of mind ; and only in proportion as man becomes enlightened by extension of new scientific knowledge, especially in physiology, will he be able to view himself in a true aspect apart from his consciousness. Science penetrates deeper than metaphysical speculation, into the nature of mental action, chiefly because metaphysics deals only with old ideas, whilst science furnishes us with new experience and therefore with new conceptions and wider evidence.

Fallacies are very prevalent, every subject of human study is liable to a very large class of errors arising from the extremely imperfect state of our knowledge, and in very few subjects is our ignorance as great as in that of mental and moral phenomena. Every different subject of study also, has, in consequence of its special peculiarities, its own peculiar class of fallacies, into which the student of it is likely to be led, unless he is previously guarded against them. In accordance with this truth, the study of man's nature, especially the mental and moral portions, is particularly liable to a class of errors arising from the circumstance, that the phenomena to be observed and the observing power are intimately connected together, each influencing and disturbing the other The obstacles to our arriving at truth in the study of mental and moral actions, are greater and more frequent, the more nearly and intimately related the phenomena to be observed and contemplated, are to the observing and contemplating faculty, or rather to the contemplative action. When the two mental actions are extremely intimate, as when attention is directed to the action of will (which is itself a conscious act of attention) undisturbed thought becomes very difficult ; and when further, the contemplative faculty attempts to contemplate itself, as when consciousness attempts to observe consciousness, in order to define it, the attempt results in almost complete failure, probably because the two actions (observing and being observed) being opposite in kind, cannot coexist at the same time in the same structure. Knowledge of the exact nature of consciousness therefore, will probably only be arrived at by indirect means, when physiological and other knowledge is sufficiently advanced.

Consciousness, when uncorrected by sufficient knowledge and inference, is a great source of error. That which we feel, we think exists whether it does or not, until the subject is correctly explained to us. The incessant and irresistible obtrusion of consciousness exercises dominion over every mind, even of our greatest thinkers, and causes disturbance and interruption in nearly every train of thought. It is largely the cause of some of our most general ideas and emotions and insensibly influences our views of man and nature. It produces true impressions as well as erroneous ones. It is a cause of the feeling that an occult spirit exists within us independent of our material structure. Combined with the almost equally persistent impression of the uniformity of nature, it largely produces the idea that the spirit within us, will live and be active for ever. And by uniting with the frequent impressions of failure of our efforts and desire for more perfect enjoyment, it largely originates the idea of everlasting happiness.

It is in accordance with modern scientific knowledge, to view the mind, not as a collection of distinct faculties, but rather as a single kind of power, like each of the physical forces, having several different modes of action ; and as that which perceives, thinks, and wills. Its oneness is shewn by its inability to be simultaneously occupied by several diverse feelings, thoughts, or volitions, and by our incapacity to think of many varied ideas at once ; the more ideas also or objects we attempt to perceive at once, the less we realize of each. In proportion as the mind is engaged upon one idea, so is it also unable to be occupied with another. Strong feelings exclude intellectual action. The mind can only execute several actions at a time, provided they have been rendered more or less automatic by habit, &c., but as all mental acts are in different degrees imperfectly automatic, and require more or less attention, and each individual mind is limited in its power, every such act withdraws a portion of attention from the more engrossing ideas. Power of mind and power of maintaining attention are nearly synonymous.

The recognised fundamental elements of mind are Receptivity and Perception of impression : Retentiveness of impression : Perception of agreement (or similarity) of impression : and Perception of difference of impression. All purely mental acts appear to be resolvable into these.

Many persons still entertain the idea that mental actions are largely independent of the natural conditions to which physiological, chemical, and physical phenomena are subject. The unscientific mind is readily beguiled by easy schemes of mental action, or simple systems of mental and moral philosophy, unaware that great abstract truths often require deep thought to discover them, or even to perceive them when discovered and published. "A false notion, which is clear and precise, will always meet a greater number of adherents in the world than a true principle which is obscure." It is not until unscientific persons have become used to advanced scientific ideas and nomenclature, and knowledge as so far progressed as to enable thinking men to illustrate those ideas freely in familiar language, that great abstract truths are believed by the public. The ordinary and simple theory of the operations of the human mind is, that they often arise without any cause, and are frequently not obedient to ordinary influences, and this idea is still entertained and promulgated even by some of our most popular ministers of truth. It is therefore necessary in order to further prove that new scientific knowledge is really a basis of mental progress, to point out a few of the chief ways in which mental action essentially depends on scientific principles, and to adduce a few instances in which other substances than brain exhibit essentially similar phenomena. To shew this however in a more satisfactory manner would require a large treatise to be written upon the subject.

The human brain and mind are evidently subject to the ordinary laws of matter and energy. Receptivity and retentiveness of impression is not only a property of brain, but of all solid matter without exception. Moser's pictures, and Chinese mirrors, the impressions on each being reproducible by warm breath, are examples of this. And as these two properties are fundamental elements of mind, they must be present in and essential to, every mental action.

All phenomena require time and all matter occupies space ; thought and brain are no exception to this. Whilst all persons say " I must have time to think," many believe that thought is instantaneous. Time is a necessary condition of all thought, and therefore of all comparison, inference, imagination, and mental analysis ; it takes time even to form an idea, or draw an inference from it, and the two cannot be formed simultaneously. Professor Donders, of Utrecht, has invented what he terms a Noëmatachograph for registering the amount of time occupied in mental processes , and by the aid of that instrument has ascertained that the time required by a man of middle age to perform a single act of simple thought is about one twenty-fifth part of a second. It has also been ascertained that the time required is longer in some persons than in others ; and longer if the subject of thought is one with which the thinker is not familiar. Mosso, by means of an instrument which he calls a Plethysmograph, has shewn, that during mental action, either in the waking state, or in dreaming, there is a greater amount of blood determined to the brain, and more during difficult than during easy mental action. These are instances of scientific research casting a light upon mental processes.

Coexistence of matter and energy is another great truth which appears to be applicable to all nature ; wherever there is matter, there is either active or stored up power ; and as particular forms of energy are in some cases most exhibited by particular kinds of substance, (as magnetism by iron), so mind is associated with living brain. As also we never see the physical powers exhibited except by material substance, so have we never yet observed mental action in a space devoid of material. The most perfect vacuum yet produced contains many millions of particles of substance in each cubic inch. Of all the countless number of scientific phenomena observed since men have been able to reliably investigate, not one has afforded us conclusive evidence of mental action entirely independent of these conditions. In accordance also with the usual truth in science, that complicated action requires complex structures; mind, being the most intricate action, is manifested by the most complicated body.

Mind, like each of the physical forces may be

viewed as a mode of energy; it is essentially dynamic; activity or change, within or without us, appears to be the original source of all our mental impressions, and the cause of their re-excitement in an act of memory. A man's mind, being continually excited by circumstances, must be active whether he will or no, and if it does not possess sufficient truthful ideas entirely to occupy it, it must be more or less occupied with erroneous ones. " We can neither feel, nor know, without a transition or change of state—and every cognition, must be viewed as in relation to some other feeling, or cognition," (Bain. Mental and Moral Science, p. 83); *i.e.* the mental effect of impressions upon us depends upon our immediately previous mental state; consciousness and perception appear to be based upon cerebral change or activity; after strong excitement of consciousness an increased amount of acid products is found in the secretions. " It is a general law of the mental constitution, more or less recognised by inquirers into the human mind, that change of impression is essential to consciousness in every form," (Bain. Emotions and Will, 3rd edi. p. 550). A sufficient degree also of such change is a necessary condition of conscious perception; it is the stronger or more rapid only of mental changes that excite our consciousness.

We perceive nearly all things by means of a difference of impression which they make upon us; by contrast. That which makes no such difference of impression, such as the great uniformities of time and space, makes no immediate impression upon us. We only know of the existence of those uniformities by inference from our perceptions of sequences or of relative difference. Although the Earth moves at the rate of 62,000 miles an hour in its orbit, consciousness does not perceive it. If also there was no error, we should be less immediately able to discern truth, without pain we should lose much of the enjoyment of pleasure. Without the contrast of imperfection we could not directly appreciate perfection.

This principle of "relativity," or of change of impression, operates both in the phenomena of dead and living matter and in those of mind ; the selenium in a photophone is kept in a state of motion or activity, not by a beam of uniform light, but only by one which changes; electrical action is excited by a relative difference of friction, of temperature, of chemical action, &c. ; chemical action also often results from a relative difference of property of two bodies. That the most inscrutable phenomenon of mind, viz., consciousness, is largely dependent upon relative physical and chemical conditions, is proved by the powerful influence which alcohol, chloroform, opium, haschish, and other substances, have in exciting or depressing it. These facts prove that excitement of consciousness or mental action depends upon precisely the same general condition, viz. : change of impression, as the excitement of some of the physical forces ; and that mind possesses a similar property to the physical forces of being changed by inequality of impression.

Whilst copious evidence is available to shew that the mind is excitable by physical causes, no more conclusive proof exists that a mental impression arises without a natural cause, than that a physical one, such as a photographic impression, arises in that way. Abundant evidence of non-creation of ideas out of nothing might be adduced ; even imagination and invention are subject to this limit, because an unlimited number of new conceptions cannot be formed from a limited number of previous ideas.

The dependence of the mind (like any other mode of energy) upon physical conditions, is further proved by the fact that the mental and moral states of a man are largely governed by sensation ; if the latter is unhealthy it makes the mind so, and it makes some difference what the part of the body is in which the sensation exists ; most commonly it is the viscera. The mind is also intimately dependent upon the physical condition of the brain, and is largely affected by the quantity and quality of the blood in that organ.

The most fundamental principle which pervades every one of the sciences, and agrees with the actions of every natural form of energy without exception, including mind, is, that of consistency or non-contradiction. No machine or scientific apparatus of any kind can perform two contradictory acts at the same time. It is both a physiological and psychological fact, that we cannot experience two contradictory sensations, nor perceive two contradictory ideas at the same instant. We can neither feel, perceive, nor observe, one thing, whilst we are feeling, perceiving, or observing, one of a contradictory nature ; nor can we perform any two contradictory acts of comparison, inference. imagination, or volition, simultaneously. As also two mental actions are often not exactly alike, or entirely harmonious, they must so far as they are really contradictory, be mutually exclusive ; and one of them must partly prevent the other, the strongest one prevailing, and this general truth is commonly though not explicitly, recognised in the maxim, that to do anything well, we must do only one thing at a time.   In accordance with the universal truth, that contradictions cannot co-exist, it is well-known that one disease frequently expels another from our frame, and the action of counter-irritants is based upon the same principle.   The fortitude of martyrs may probably be explained by this power of one set of ideas and feelings to exclude another, and the facts of mental physiology afford plenty of other examples.

It is probably because we cannot simultaneously perform two contradictory actions, that we cannot contemplate consciousness, or think of an idea and at the same time think of that act of thought.   In accordance with this, even Newton, and other great geniuses, have been unable to accurately describe the mental processes by means of which they arrived at their most difficult results.   In consequence also of this, we cannot define consciousness, and are often unable to directly observe or analyse our mental actions, especially those of a very abstruse or complex kind. Much of the knowledge of the operations of our mind, we are therefore obliged to obtain by indirect means; by analogies, and inferences from the phenomena of nature, &c., and in this way our knowledge of mental action largely depends upon our acquaintance with physical and chemical science, and can only advance as it advances. To clearly understand one subject we are often obliged to study several others. Ignorance of science in general, and of cerebral physiology in particular, is the chief obstacle to our acquiring a more accurate knowledge of mind.

Next to consistency, the great principle of causation constitutes the most essential part of all natural truth, and to deny the operation of this principle in particular cases of mental action, simply because we, with our very finite powers, cannot in the extremely imperfect state of our knowledge; yet fully explain some of the most difficult, complex, transient, and ever-changing phenomena of will and consciousness, is contrary to the most weighty evidence. "The Will" is a conscious mental effort to effect an object, the idea of which is already in the mind, and being a mental "effort" it absorbs the mind and thereby incapacitates it at the moment from observing its own action.

If any phenomenon (such as mental action) is essentially dependent upon another, it must be connected with it in a never-failing or indissoluble manner, so that when the one occurs the other is always present, otherwise it would not be essentially dependent. The only known connections of this kind are those causation and continuity of phenomena, according to which every phenomenon has a cause, and all phenomena are indissolubly connected in endless series. The evidence of the truth of these principles is so vast, that even all mankind thinking through all ages, and after having made an almost infinite number of definite experiments and observations, have never yet met with a single well verified instance of their failure; and we are therefore justified in inferring that they are universal. There are however instances in the physical and chemical sciences, as well as in mental action, where the dependence of phenomena upon those principles is not very apparent, and has not yet been sufficiently proved, but it is probably in consequence of our imperfect knowledge and limited faculties, that we are unable as yet to fully trace such dependence. The history of science, abundantly proves that we should not assume that a phenomenon arises without a natural cause, simply for the reason that it is very difficult to trace its origin, but wait patiently for more knowledge respecting it. It is unphilosophic and contrary to reason to attribute to occult agencies, effects which may be explicable by ordinary causes, or to refuse to believe in more abstruse causes where the assumption of simple ones is contradicted by some of the evidence.

The principle of causation forms the basis of many minor ones, such as selection, evolution, differentiation, &c. Plurality of causes also is a very common circumstance in all the sciences, and especially in concrete phenomena, and in the complex ones of animal life ; the arrival of a ship for example at a distant port, is a result of many conditions. Similarly with most of our mental actions, they are compounds of feeling and intellect, and produced by many causes, such as hereditary tendency, acquired habit, internal and external mental excitants, dogmatic belief, knowledge of empirical rules, and occasionally of verified principles. Several of these causes also frequently conspire to produce a single idea or decision.

Various general principles of lesser magnitude arise from the combined action of two or more of the greater ones, and these also appear to operate in mental actions as well as in physical ones. Thus by the combined influence of causation and of the principle that every phenomenon occupies time, "effects often lag behind their causes ; " and in some cases during a long period. The greatest heat of summer for example usually occurs several weeks after midsummer. The mental effects of early mistakes are often not fully experienced until old age. The decline of a nation also follows a long time behind the period of action of the chief causes which produce it.

Although effects are indissolubly connected with their causes, they frequently do not occur in an active form until a long period after them. In such cases they are stored up in what is termed a potential or latent state, ready for liberation at a future occasion, when the suitable conditions are present; the storage of chemical power in gunpowder, of solar heat in coal, and its subsequent liberation in our fires, are suitable examples. The principle of deferred activity and storing up of power, occurs also in vital and mental phenomena; potential heat is stored up in our food, and is afterwards evolved by oxidation in our tissues. Muscular power is stored up during sleep, ready to be evolved during labour. The storage also of cerebral impressions, and cerebral energy, ready to call forth ideas, and thereby powerful emotions, by the exciting action of memory, may also be viewed as an instance of similar kind belonging to mental phenomena. A new and striking instance of the storage of energy has been shewn in Faure's improved form of secondary voltaic battery, in which the most powerful voltaic current may be (at least practically) stored up (in a box containing lead plates immersed in dilute sulphuric acid) and conveyed to a distance with little loss, and then liberated.

Exciting causes operate very extensively in mental actions as well as in physical ones, a mere look or word from an eloquent speaker will excite the passions and liberate the muscular power of a multitude. Every part of the human body, especially the muscles and nerve centres, is a store-house of power always ready to be set free by the slightest suitable causes; this is strongly illustrated in the irrepressible activity of children, and in the excitable passions of young men and women. The more immediate cause of this power is the oxidation of assimilated food ; and the source of power in the food is the heat of the Sun stored up in the plants and animals they have eaten.

The subsequent liberation of power under the influence, often of very slight causes, long after the original cause has ceased to act, has led us to conclude erroneously that causes are not always proportional to effects. Proportionality of effect to cause appears to be universal ; it probably operates in mental as well as in physical actions, our faith in education as a means of intelligence is based upon this ; the more complete the education of a particular individual, the greater usually is his degree of intelligence. Proportionality of cause to effect is apparently disobeyed not only in physical but also in mental phenomena. Throughout the whole realm of nature, minute circumstances often act as exciting, deflecting, and guiding causes, and contribute to the production of apparently disproportionate effects. Thus a spark will discharge the largest cannon ; a touch determine the most distant electric signal ; a word or look, excite the strongest emotions ; the little change of position of a railway point will direct a train either to distant North or South; the minute change of contact of the telegraph switch, will determine the signal to places wide asunder ; one false idea also at a critical moment will often lead a man or woman to ruin ; and in all these classes of cases, whilst trifling causes *appear* to produce great effects ; the real causes are the stored up latent powers set free or directed. It is astonishing how small a circumstance will excite an idea, and deflect the entire current of our thoughts ; and it is equally surprising what great physical and chemical effects are often started by most minute exciting or deflecting conditions ; the explosion of seven tons of dynamite at Hell-gate, near New York, by the pressure of a child's finger closing an electric circuit is a suitable example.

Every phenomenon therefore whether physical or mental, is probably connected in an indissoluble manner with some preceding phenomenon, either immediately in point of time, or remotely through some static condition, usually that of stored up power. In this sense the great principle of continuity of phenomena appears to be universal, and the present state of the Universe is said to implicitly or potentially contain all the future states of the Universe. Mind also in this way, like each of the physical forces, often acts as a link in an endless chain of causes and effects, and is connected with non-mental phenomena in accordance with the great principles of science.

Science has demonstrated what has been termed the "Convertibility of Forces," or, that when one form of energy disappears, another form (or forms) of energy, and in precisely equivalent amount, is produced in its stead, either in a latent or active state. The equivalent quantities of the various forms of

energy have also been discovered by actual experiment and measurement. A pound weight falling
through 772 feet gives forth as much energy as would
(in the form of heat) raise the temperature of one
pound of water one Fahrenheit degree. We know
that so much mechanical power is equal also to so
much electric current, chemical action, &c., and a
large amount of evidence exists to show that these
transformations of energy occur in all the organs of
living creatures, and in obedience to the law of their
equivalents. How far mental power is a "mode of
energy" transformable, and obedient to the laws of
equivalence, are interesting questions for future
research.

The mechanical principle of action and reaction is
another which can be traced in mental as well as in
physical phenomena. Mental excitement is often
succeeded by mental depression, "after pleasure
follows pain." The power of mental self-guidance
and self-education is largely dependent upon the two
well known scientific principles of latent energy, and
action and reaction. We are able to liberate energy,
not only in cases where it will influence inanimate
matter but also ourselves. The principle of self-
guidance is not restricted to living creatures, nor is
self-regulation limited to mental power. The principle of self-regulation operates in clocks, watches,
musical boxes, the governors of steam engines, water
regulators, gas regulators, &c., &c., and upon an
immense scale in the movements of the heavenly

bodies. With the electric locomotive, the greater the load it has to draw, or the steeper the incline it has to ascend, the more strongly does it exert its strength, up to the full limits of its power. Neither in physical nor in mental actions can a body or force usually act directly upon itself to change its state whether of activity or rest. In both classes of cases however we meet with plenty of instances where, a body by an almost imperceptible expenditure of energy on its own part either alters some surrounding conditions, or excites a powerful liberation of energy in another body which then reacts upon it to change its state. In this way the action of clock-work in the self-exploding apparatus of a torpedo liberates at a particular moment a spring, and causes an explosion which destroys the apparatus. Similarly, whilst a man, in many cases, is unable to *directly* alter his mental state, to increase or diminish his mental activity, to cause sleep, &c., he is able *indirectly* to change his mental condition by drinking stimulants or by adopting means of self-education ; and to induce sleep by means of opium, suitable exercise, &c.

The principles of indestructibility or conservation of matter and energy, flow from the preceding ones, and are exhibited in mental actions as well as in physical ones. Whilst the universal experience of mankind has not yet afforded us a single well verified instance of actual creation or destruction of matter or energy, it has supplied us with plenty of examples of apparent destruction and creation of each of them.

But scientific knowledge corrects the uncertain testimony of consciousness ; whilst we see coal burn and be apparently destroyed, science proves to us that the elements composing it remain undiminished. We observe also that the heat of the fire dissipates and is apparently lost for ever ; but science again proves that it is either stored up in the latent state, ready to be again liberated at a future time, or else converted into other forms of energy. A given atom of matter or a portion of energy, therefore, to the best of our knowledge, continues and persists for ever. As we cannot either create or destroy matter so also can we not create or annihilate energy, and this truth probably holds good with regard to mental as well as to physical and chemical power. Great changes of state in bodies (as in the combustion of wood &c.) have led us to erroneously think that the substances are destroyed; and great apparent differences of property, such as those of diamond and charcoal, have led us similarly to conclude that they are entirely distinct and independent of each other when they are not.

As the cerebrum of man is composed of matter, and during excitement, its parts are active, we might confidently predict that its particles obey the First law of Motion, viz. : that a body in a state of rest or motion will continue in that state of rest or motion until some cause arise to prevent it. So it has been found that the action of the vital and mental forces have a degree of persistence, like the physical ones. It has been experimentally found that portions of living bone transplanted to fleshy parts of animals where there was no bone, continued to grow for a time by a life of their own, and increased by formation of additional bone, like a crystal grows in its medium ; but after a time they diminished and disappeared. In a similar manner we are all of us aware of the persistency of ideas, even in opposition to the will, after the cause of them has been removed. Sometimes we cannot retain an idea because of the persistence of others ; and at other times we cannot get rid of one, for a similar reason. Our mental habits also have often very great persistence.

The principle of heredity may be viewed as a result of the First law of Motion, and appears as Persistency of state, either of structure, form, or mode of action. It appears both in inanimate bodies, living structures, and in mental phenomena; in the latter, as hereditary mental peculiarities. The principle of Persistency of structure and Heredity of form and property, during repeated or even continual dissolution and aggregation of a material substance, is more or less manifest nearly throughout the whole of nature. In the formation of crystals it is clearly seen ; each crystallizable substance will only grow into its own shape or shapes ; each particle of common salt, during an endless series of successive solutions and aggregations into the solid state, always forms a more or less perfect cube ; that of silica a hexagon ; and so on throughout the entire series of thousands of different crystalline bodies. As each form of crystal only produces crystals of like form and property (or at most in certain cases a limited number of modified forms, as in the instance of calcic carbonate, &c.) so also each seed, both of animals and vegetables, only produces its own particular essential shape and collection of functions. The same principle shews itself in the transmission of particular types of disease, and of eccentricities of organization, from one generation to another of animals. Peculiar malformations of body and characteristics of mind often persist in families from generation to generation. This persistency or heredity of structure and of property is not limited to solid bodies, but exists also in liquids: "The effect of vaccine virus upon the liquid blood, in producing a permanent and organic change in its constitution and character, which continues to exercise a protective influence against small-pox, in the great mass of cases, through a long life, during which time the blood must have undergone, many thousands, if not millions of changes and modifications." (F. Winslow. "Obscure diseases of Brain and Mind," page 432). The same persistency of structure and property of structure, has even been detected in vapours ; the vapour of red iodide of mercury for example, deposits only crystals of red iodide, whilst that of the yellow deposits only yellow (see Gmelin's Handbook of Chemistry, vol. 1, p. 100.)

We often appear to mentally select when we only yield to causes acting upon us, *i.e.*, to the strongest influence or motive. That " self-preservation is the first law of nature," is not only true of living creatures, but largely also of dead substances. Inanimate as well as animate matter, appears to usually select what is good for itself. Apparent selection, which is manifested in the phenomena of instinct, is exhibited not only by brain, but by all material substances. Acids appear to select bases, North magnetism rejects North and prefers South magnetism. Also if a piece of zinc is put into a mixed solution of the nitrates of silver, magnesium, calcium, strontium, barium, lithium, sodium, potassium, and rubidium, it will select the silver only with which to form a " metallic tree," and reject all the other metals. Everything which aggregates or grows to a definite shape, appears to select its material ; if a crystal of a particular salt is placed in a mixture of saturated solutions of different salts, it will only select and assimilate to itself suitable material, either particles of the same composition as itself, or those which are isomorphous with it, *i.e.* belonging to the same crystalline system. In living bodies also, the same principle operates ; Living tissues, whether of animals or vegetables, usually select from their nutrient fluids, and assimilate, particles only of those kinds of matter which are suitable for their structure ; in this way, a bone assimilates lime and phosphoric acid from the multitude of different substances conveyed to it by the blood. And in all these cases, the selecting material appears to act as if it possessed the powers of instinct, perception, comparison, judgment, and volition. The act of self-repair is clearly connected with this, and is not limited to living structures ; Sir David Brewster observed that if a portion of the surface of a perfect crystal of alum is very slightly abraded by dissolving a film from it, and the crystal be then immersed during a very brief period in a saturated solution of alum, the abraded portion repairs itself. The subjects of "malformation of crystals," and "diseases of crystals" have been scientifically investigated. The power of selection (or rather of apparent selection) is no doubt a result of the combined action of causation and of the inherent properties of bodies, and depends, like consciousness, upon difference of impression, the strongest suitable influence determining. If apparent selection can thus be performed by inanimate matter, we should not, except for a very sufficient reason, assume the existence in living creatures, of a special occult power to perform the same function. In the selection of ideas also the intellect acts according to the purely scientific method.

We frequently appear to mentally adapt ourselves to particular circumstances when we are really determined by causes ; and this apparent adaptation is also seen in ordinary physical and chemical phenomena The course of a river for instance, adapts itself to the configuration of the country through which it flows, and if it cannot pass wholly by one channel, as in seasons of flood, or on occasions of accidental obstruction, it travels through several ; and a similar result occurs with the flow of the blood when an artery is tied or becomes obstructed. A plant when growing in a dark recess, bends itself towards the light as if it preferred light ; and its roots adapt themselves to the forms of existing obstacles. A decapitated frog jumps away from a source of irritation, as if he still possessed sensation, volition, and choice. A man seeking his way through a crowd avoids the course in which the the throng is densest. The human mind also, chooses as it were, the easiest way of solving a problem, and usually adapts itself to altered circumstances.

The principle of evolution also operates both in physical and mental actions, and is a result of that of causation. Complexity of structure and function is evolved out of simplicity of composition and property by plurality of causes and conditions. For instance, many complex forms of crystals of ice are produced from water. Calcspar crystallizes in more than one hundred varieties of form, (all derived from an obtuse rhombohedron) under the influence of a number of slightly different conditions of temperature, impurities in the solution, &c. The most complex bodies are evolved out of the simplest, the bodily frame of man himself (and that of other animals) is constructed of less than twenty of the elementary substances. The same simple substances are capable of yielding very different and more complex bodies under different conditions ; thousands of different chemical compouuds are composed of hydrogen, oxygen, and carbon only. In the development of living forms from ova, the ultimate form produced does not exist in the germ, any more than a crystal exists in its solution, but is a natural consequence of the forces acting in and upon the germ, like the cubical form of a crystal of common salt is a result of the forces acting in its constituents under the conditions of its environment, especially those of pressure and temperature. The extent to, and manner in which, the force and principle operate, depend upon the material substance, and its conditions internal and external.

It is a common circumstance, both in physical, mental, and moral subjects, for the apparent to be the very opposite of the real. This general truth has been repeatedly illustrated in an incidental manner in this book, and need not be much further elucidated. Phenomena are none the less real, however, because they are not readily manifest ; our earth is as much tied to the sun by the invisible power of gravity, as if it was attached to it by visible material chains. Mistaking the apparent for the real, largely explains the persistency of certain beliefs, and why it is that persons unacquainted with science, cling to self-deception, and resist some of the most firmly established truths. The more evident but untrue explanation is believed, whilst the less apparent but true one is rejected. It is the chief cause of the belief that " the will is a supernatural power." To a scientific man, however, apparent contradictions are not unfrequently a sign of truth ; too accurate results sometimes indicate that they have been artificially made to appear correct.

Sympathetic action or propagation of similar influence by immediate impulse, is a property of all the natural forms of energy, as well as of mind. Similar actions are propagated thus in all kinds of dead substances, as well as in the living brain. Matter is sympathetic to sound in the phenomena of singing-flames, and a vibrating string responds to a particular note in obedience to well-known laws. Iodide of nitrogen may be caused to explode by the influence of a particular note from a fiddle. In the phenomena of light, with a spectroscope, a luminous gas is sympathetic with, and emits and aborbs, only particular kinds of luminous rays. In chemical action also, combustion excites combustion, ferment excites ferment, infection communicates infection, and the similar chemical change is transmitted from molecule to molecule. Mental excitement and disease in one person, often excite similar phenomena in another, as is seen in "religious revivals," and well-known epidemics, such as the "dancing mania," "preaching epidemics," the "leaping ague," the "mewing contagion," etc., etc., (See "Epidemics of the Middle Ages," by Hecker; Sydenham Society publications; also Carpenter's Mental Physiology, p. 312.) Like excites like in the actions of each of the forces of nature ; both in physical, chemical, and mental action, the kind of impulse transmitted is similar, unless conditions exist which transmute it. Dynamite, started into combustion by a flame, burns slowly away; but when started by a detonating substance, detonates violently. Guthrie has also discovered that if a melted cryohydrate (*e. g.* a chilled saturated aqueous solution of a solid salt) is cooled to a certain greater extent, it will not solidify—nothing separates out, although the solution is four or five degrees below its proper solidifying point. If a little crystal of ice be then thrown into it, nothing separates but ice, which comes to the surface. If we throw in a little anhydrous salt, nothing but the anhydrous salt separates out, and that sinks to the bottom. But if we throw into it a crystal of a previous crop of cryohydrate, then nothing but the cryohydrate separates. In this case also, like evidently excites like only, in obedience to physical laws. (Addresses, Science Conferences; South Kensington Museum, 1876; Vol. 2, p. 108). Even two clocks, when hung near each other, against a board or surface which readily transmits vibrations, have been known to exhibit, by synchronous action, an apparent sympathy, which changed their rate of going.

Periodical phenomena, also, brought about by definite causes, occur in mental, as well as in physical phenomena. In the former we have the phenomena of sleep, and in the latter, definite causes produce summer and winter, day and night, the tides, cycles of solar spots, maxima and minima of magnetic intensity, etc., etc.

Conversely to the manifestation of the principles of

inanimate matter by living bodies and in mental action, so have modern inventions demonstrated the possibility of the performance by inanimate substances and apparatuses, of the functions, not only of our bodily organs, (as of locomotion by the steam engine,) but also of our senses and intellect, and in some cases, to a degree far surpassing unaided human power. Apparatus, sensitive to sound, have been constructed, as in the microphone and singing-flames ; others capable of reproducing articulate speech, as in the phonograph, telephone, etc.; others again sensitive to light, as in the production of visible images by photography, and reproducing them at a distance through wires by means of the photophone; the power of indicating or foretelling future events has also been embodied in instruments called "tide predictors," and that of evolving inferences has been shewn in Jevon's " Logical machine."

The various facts mentioned in this chapter prove that mind agrees with the various forms of physical energy in many essential points, and obeys many of the same laws or principles. Examine whatever general phenomena of the mind we may, we can always detect some apparent or real connection of them with the great principles of inorganic nature ; and in order to prove the dependence of them upon the great principles of science it is not necessary to show that all such actions are subject to those principles. Until the whole is explained however, there will always remain mysterious phenomena to cavil about.

MORAL PROGRESS.

At the present time few competent persons have largely investigated the fundamental relations of morality to Physical Science, consequently moral actions are usually considered not to have a scientific basis, and the doctrine is still extensively taught that some moral phenomena are essentially supernatural.

As new scientific knowledge has increased, belief in witchcraft, sorcery, demonology, exorcism, evil influences and omens, unseen spirits, a God of evil, supernatural and occult powers, supernatural sources of strange diseases, evil presages from comets and eclipses, fetishism, worship of images and of the Sun, a belief that the Earth is the chief body in the Universe, that man is the "Lord of Creation," &c., &c., have largely passed away, and beliefs more consistent with facts and with true inferences drawn from them, have taken their place. Belief in the supernatural nature of the human will however is still largely retained. A writer on morality, says—"In every genuine volitition we have a phenomenon not law-determined, law-regulated, and law-explained."* A popular expositor of religion says—"The phenomena of the human soul are essentially different from the phenomena with which the student of science is most familiar, and must be investigated on other principles and by other methods." "The voluntary activity of man lies beyond the limits of science." "Every language man has ever spoken—no matter how per-

* "Wish and Will," by L. Turner, M.A.

fect or how rude—the literature of the ancient and
the modern world, the indestructible instincts of the
human soul, the testimony of consciousness, unite
to affirm that the human will is independent of
natural law." " The will is a supernatural power."
"I myself am not under the dominion of natural law;"
" my moral life is essentially a supernatural thing."
"As soon as you approach the intellectual and
moral life of man, you enter a region in which you
have to do with a new order of facts."   &c.*

Morality, the subject of duty, or of right and wrong-
doing in conscious creatures, is usually considered to
relate only to those actions over which a man has or
might have had control, and which it was his duty
either to perform or avoid, and not to those which
are entirely beyond his influence, it is therefore
essentially dependent upon the power of selection or
choosing.   As then all moral actions require voluntary
choice between right and wrong, and every act of
choice is a mental one of comparison of two or more
things, all moral actions are mental ones. We cannot
compare things which have not made any mental
impression upon us. We know further, and the
evidence already given proves, that mental actions
are intimately dependent upon the principles of nature
operating within and around us. If then all acts of
morality (or immorality) are mental ones, and if all
mental actions are intimately dependent upon the
great principles of nature ; then all acts of morality

* " The Mutual Relations of Physical Science and Religious Faith."

are dependent upon those principles. Morality also cannot be properly understood without a knowledge of various sciences, especially biology, because it relates to human creatures, all of whom are morally affected by the various forces and substances belonging to the physical and chemical sciences.

Having shown that moral actions are mental ones, and adduced evidence to prove that mental actions are largely subject to scientific principles;—it follows as a necessary consequence, that moral actions also largely obey those principles, and I need not repeat that evidence.

The extension of scientific knowledge conduces in a very general way to moral progress, by diffusing the " scientific spirit," increasing our love of truth, facilitating the attainment of greater certainty and accuracy, enabling us to more perfectly avoid error, reducing our ignorance, dispelling superstition, inculcating obedience to law, diminishing difference of opinion and thereby lessening strife, conducing to humility, to greater economy of means, to increased cleanliness, &c, &c. Scientific research also, by disclosing to man his true position in nature, enables him to act in harmony therewith, and thus increase his morality and general happiness.

Knowledge is as free as the air, once diffused it becomes impressed upon the brains of men and cannot be easily destroyed or restrained; and the greatest moral effects of science are cosmopolitan ones. Inventions based upon new scientific truths are gradually breaking down the barriers between the various nations of the Earth, and infusing common interests amongst all mankind. Nothing is uniting the sympathies of different nations, increasing the friendly feelings between them, and diminishing the probability of war, more than the increasing facilities of communication brought about in a great measure by the devolopments of science and art; more particularly by ocean steam navigation, rapid postal communication, and the telegraph, (see p. 51). At the present time there are about six Atlantic telegraph cables in use, and an almost daily service of passenger steam ships across that ocean. The use of inventions based upon scientific discovery has aided moral progress in various ways. All inventions are made with the object of supplying some real or supposed want, and nearly everything which supplies a common want, conduces to contentment and happiness and the general progress of mankind. No one can possibly measure or estimate the advantage of the inventions of writing and printing, in helping men to avoid quarrels, to settle differences of opinion, to sympathise with suffering, to give advice: &c. Similar moral functions are also performed by the electric telegraph, and a few specimens of some of the messages sent through the wire would clearly illustrate this fact. Great moral progress has also resulted from cheap daily intelligence, collected largely with the aid of the telegraph ; and of cheap books produced by means of the steam engine. It is estimated that 250 millions of copies of newspapers are yearly published in Great Britain. The Bible and Religious Tract Societies could hardly have existed had not the properties of the ingredients of ink been discovered. The present multiplicity of testaments, prayer-books, hymn-books, &c., has also been rendered possible by the invention of printing, As darkness is favourable to crime, so the invention of gaslight has conduced to morality. The numerous sources of intellectual and moral enjoyment, developed by inventions based upon scientific discovery, have attracted mankind from more sensual and less moral amusements, and the invention of the piano-forte has operated largely in a similar manner.

In many respects, the poor man of to-day can command social comforts, conveniences, and pleasures, which an emperor could not in former times. Who can estimate the amount of beneficial moral influences of an indirect kind obtained by means of modern science ? The relief from pain by chloroform and other new medicines, the diminution of domestic toil by the sewing machine ; the increased health and pleasure obtained by access to the country and sea-side by means of railways ; the diminution of anxiety resulting from more speedy conveyance of letters, and especially of messages by telegraphs, the increased pleasure of life resulting from being surrounded by objects of beauty multiplied cheaply by means of scientific processes, such as photography ?

The human mind cannot greatly resist impression, the various effects of scientific research necessarily produce an influence upon it. Whilst great deep-seated truths make a powerful impression on the minds of philosophers, the great practical effects of science in inventions, &c., profoundly impress the mass of mankind. One of the chief influences of the discovery of important scientific truths and of their practical application in some wonderful way, such as in the telescope and microscope, phosphorus matches, photography, electro-plating, the electric light, the spectroscope, microphone, telephone, &c , is to produce a profound and wide-spread impression of the existence of a great and mysterious influence, which produces (or enables us to produce) such striking effects.

Whilst also the novelty of the practical effects of new scientific truths in inventions, astonish persons in general ; the definiteness of scientific phenomena, and the certainty with which they may be reproduced convince all competent persons who examine them, that they are rigidly subject to definite laws. In this way the antique belief that natural phenomena are produced by supernatural agencies, is gradually being abandoned, and the more moral conviction of the omnipresence and universality of law has been largely established in its stead. Every new scientific fact and invention thus becomes a new proof of the universality of law. Belief in the supernatural has diminished in proportion as scientific knowledge has advanced ; instead of natural phenomena being erroneously ascribed to demons, spirits, supernatural powers, and occult causes, they have been proved to be results of natural powers, acting in accordance with known principles. Assertions which have been made, that "the will is a supernatural power, independent of natural law" &c., are not supported by evidence at all equal in cogency to that in proof of the statement, that our mental and moral powers as a whole act in accordance with the great principles of science.

That moral phenomena, like those of the physical sciences, are capable of being made the subject of experiment, observation, comparison, analysis, and inference, is very manifest. Every case of bribery may be viewed as an experiment in morality. The very common case where an employer tests the honesty of a servant by some contrivance, is also a trial of a similar kind. The dependence of the moral powers upon scientific conditions, is clearly seen in the influences of intoxicating drinks. A mere natural substance could not possibly overcome the influence of a power which exists entirely independent of it ; *i.e.*, a " supernatural " one. Even the greatest believer in the "supernatural " power of the human will, deplores the serious injury which the abuse of alcoholic liquors produces upon mankind, rendering the will powerless, and debasing the moral sentiments. The effects of opium, haschisch, &c., are other examples. A vast number of experiments remain to be made of the effects of drugs and organic compounds, both solid, liquid and gaseous, upon moral actions ;

which will probably prove a still greater degree of dependence of those actions upon purely physicial and chemical conditions.

The "order of facts" in the subject of morality requires precisely similar mental treatment to those to which scientific investigation has been already applied with such great success, and which include all phenomena admitting of observation, comparison, analysis and inference ; and not only those in which we are able, but also those in which we are not able to produce by means of experiment, the phenomena to be observed, such as those of astronomy and geology. Different subjects also are experimental in different degrees, physical science is more experimental, physiology is more observational ; morality is partly experimental, and therefore capable of reduction to scientific system by means of our intellectual powers.

In consequence of the essential nature of truth being the same in all subjects, and of the fundamental processes of mental action in the determination of truth being also alike in all, the essential modes of arriving at and detecting moral truth are the same as those employed in research in the physical sciences. We possess therefore no special faculty, call it "conscience," or what we may, by which we are enabled to infallibly arrive at truth in moral questions. What is right and good, and what is wrong and evil, are determined by precisely the same general means as what is true ; our much vaunted consciousness alone does not infallibly tell us ; reason alone, acting upon the evidence, is the final arbiter in any doubtful or disputed case. The truth of moral questions must be examined and tested by precisely the same mental faculties and processes as those employed in physical science, viz :—by the faculties of perception, observation, comparison, and inference, acting upon the whole of the evidence ; and by the processes of observing facts, comparing them, inferring conclusions ; by analysing and cross-examining the evidence in every possible way, and extracting from it the largest amount of consistent knowledge.

Although we cannot detect moral truth by any other than intellectual processes, we may however arrive at correct moral conduct in two ways, viz :— either blindly or intelligently. We arrive at it blindly or automatically by the process of trusting to our inherited and acquire tendencies and dogmatic beliefs ; and intelligently by the conscious use of our knowledge and intellectual powers ; and each of these methods has its advantages. The former process' being an empirical one, is very uncertain and cannot be employed for the judical detection of truth, or the certain discrimination of it from error, it has however to be trusted to in all cases where we are deficient in knowledge, or have not time for investigation. Truthful ideas and correct conduct also, which at first require the exercise of considerable intellect and much self-discipline, in order to arrive at them, become by habit so completly converted into acquired tendences as to be automatic. It is not improbable that many of our truthful ideas and correct tendencies were originally arrived at by intellectual processes; and have become incorporated into our mental and physical structure by habit, education, and inheritance.

The scientific basis of morality is further shewn and essentially proved by the fact that the fundamental rules of morality are dependent upon scientific principles. According to Dr. Clarke, the two fundamental "rules of righteousness" which regulate our moral conduct are, first, that we should do unto another what we would, under like circumstances, have him do unto us; and second, that we should constantly endeavour to promote to the utmost of our power, the welfare and happiness of all men (to the latter might well be added, the welfare of all sentient creatures). The first of these rules is essentially dependent upon the scientific principle of causation, viz :---that the same cause, acting under the same circumstances, always produces the same effect, if what we did for another person under like circumstances might produce a different effect to what it would when done for ourself, the rule could not be depended upon and would be of no use. The second also agrees with the great principles of science, for the more we obey those principles, the more do we really "promote the happiness and welfare of all men." The first of these rules however in the form usually stated, is incomplete, because it does not provide for the circumstance, that many persons desire to have done unto themselves, not that which is most right, and really most for their welfare and that of mankind in general, but that which would most please them. The desire of immediate pleasure or consolation is greater than the love of truth in nearly all men, and this is connected with another fact, viz :—that persons unacquainted with the great pr.nciples of science, have not the advantage of the moral sustaining power of those principles, and are compelled in circumstances of trial to seek extraneous mental relief.

The desire to do right is not the primary source of morality ; there must be a cause for that desire, and this fact also shews that moral phenomena are dependent upon the scientific principle of causation. We can also mueh better understand a subject, especially a complex one like that of morals, when we can co ordinate its facts in a scientific manner, by referring them to some general principle which governs or includes them. Referring moral actions to a verified scientific principle, is more satisfactory than referring them to a less definite source such as " conscience " ; the " testimony of consciousness " ; or " the indestructible instincts of the human soul," because a principle affords a more consistent explanation than a dogmatic idea. The fact also that the discoveries of science usually precede the developments of the moral advantages of science to mankind, is in harmony with the general truth that effects follow their causes, and with the conclusion that moral rules and moral progress have a scientific basis.

In a general way, the influence of science upon moral progress is connected with what has been termed, "the scientific spirit." This characteristic consists mainly of an intense love of truth, a desire to acquire new knowledge, to arrive at certainty and accuracy ; also an obedience to law in general, and a consequent philosophic resignation to inevitable ills. Science inculcates these qualities, and it is well known that scientific discoverers have usually been highly moral persons, truthful, accurate, law abiding, patient, persevering, temperate, &c. On the other hand, the most lawless persons are usually those who are most ignorant of the great laws which govern their actions, who over-estimate human power and ability, and are impelled by ill-regulated enthusiasm or feeling.

Belief in and obedience to law, being a fundamental moral quality, is in its turn the source of other moral qualities of less importance ; for instance, it tends to produce calmness, resignation, contentment, patience, submission to the inevitable, &c. No man can be highly moral who disobeys the great principles of nature. We may however obey those laws either intelligently, by acquiring a scientific knowledge of them ; empirically, by obeying rules framed in accordance with them; or blindly, by obeying dogmas which happen to agree with them. Those who do not understand laws cannot of course intelligently obey them, and those who most disobey them, consist nearly wholly of those who do not understand them.

Superstition, ignorance of natural law, and a belief in occult powers, encourages lawlessness, injures the moral sentiments, and is often attended by bigotry, associated with strife, schism, and sectarian dispute.

Probably the greatest influence which scientific discovery has had upon the moral progress of mankind, has been by inculcating an intelligent love of truth on account of its own intrinsic goodness ; in this respect it stands pre-eminent. Love of truth is a fundamental virtue because it is the basis of many smaller ones. It is more virtuous, also, to pursue truth on account of its own intrinsic and unqualified goodness in all respects, than for any narrow extrinsic quality, such as the personal pleasure or utility it may afford, or on account of any personal gain or loss, reward or punishment, which may result from pursuing or neglecting it. In the present imperfect state of civilization however, the great bulk of mankind unavoidably employ less noble, as well as the noblest motives, as a means of improvement. Most men can only be moved to do right by means of inferior motives, one of the most effectual of which in a commercial nation is " small investments, large profits, and quick returns ;" an expectation of great reward in return for small self-sacrifice.

The discovery and dissemination of verified scientific knowledge is a purer kind of occupation than the promulgation of any kind of dogma, because the statements of verified science are usually capable of demonstration, whilst those of doctrine, being often contradictory, may, or may not be true; mere affirmation also, when not based upon proof, is often dangerous to morality. In dogmatic subjects a man may tell untruths with impunity, because no one can disprove or convict him ; but in demonstrable ones, a man dare not utter falsehoods, because others will prove his statements to be erroneous. It is demonstration rather than doctrine that is of divine origin. A man also who practises scientific research is largely compelled to adopt the most truthful views of nature, in order to enable him to make discoveries,

Real science is largely independent of opinion or faith. Whether we believe or not that a piece of clean iron immersed in a mixture of oil of vitriol and water, evolves hydrogen gas, the fact itself remains unaltered. It is a great and glorious circumstance for mankind, that human progress depends essentially upon a knowledge of new verified truth. As verified experimental knowledge can only come from the great source of all that is good, to doubt the value of new demonstrable truth, is practical atheism. Those also who systematically investigate sources of verifiable truth, are much more likely to ultimately arrive at the fountain of all truth, than those who employ unsystematic methods, or prefer unproved beliefs to verified knowledge.

Another of the most powerful ways in which scientific discovery has promoted moral progress has been an indirect one, viz., by diminishing ignorance. Deficiency of knowledge is the parent of a vast amount of evil and failure. "There is no instance on record of an active ignorant man, who, having good intentions, and supreme power to enforce them, has not done far more evil than good." (Buckle, "History of Civilization," vol. ɪ, p. 167). Ignorance largely precludes happiness, and intelligence is an indispensable condition of the highest morality. There are plenty of difficult positions in life in which the desire to do right is not alone sufficient, we must intelligently know what is the right course to pursue. We are all of us ignorant in different degrees, and must be content in many matters to walk by faith until we can walk by sight, and to act according to rule and precept until we have discovered general principles to guide us :—blind dogmatic morality and "rule of thumb" method is vastly better than none, and has rendered great services to mankind. Whether comforting doctrines are true or not, the great bulk of mankind prefer them because they afford immediate relief ; and whether they be erroneous or truthful, men will be benefited by them and continue to believe in them, until their minds are sufficiently advanced to receive a knowledge of verified principles. Rules of morality however, when presented to us with a basis of demonstrable truth, come with a degree of divine authority, and possess greater claim to our observance, than the same rules presented to us as empirical or dogmatic statements only.

In proportion to our ignorance the more we dislike to be apprised of our defects and the more inclined are we to continue uninformed ; because the less intelligent we are the less are we able to perceive the evil effects of our blindness or the advantages of knowledge. As also the present state of civilzation is very imperfect, and unsolved problems exist in all directions, ignorance and all its evil consequences are extremely prevalent. It causes the great mass of mankind to neglect better objects for the sake of money. It indirectly constrains lawyers to neglect moral evidence. It induces medical men to withold truth from ignorant patients. It causes ministers of religion to prefer doctrine to demonstration. It would therefore be comparatively easy to compose lists of our moral deficiences, and of improvements urgently needed in morality, far more extensive than the very incomplete ones of our material shortcomings already given (see pp. 68 to 78). To enumerate however the imperfections in the moral conduct of mankind, the frauds in trades, the undue advantage taken of the defenceless, the deceit and empiricism in professions, the professional trading on human weakness, the cruelty of field sports, the hollow motives of social, political, and religious life, the propagating as infallible truth, doctrines which may be fallacious, is not the object of this Chapter ; but rather to make clear the fact, that the extension of the domain of verified truth by means of scientific research is highly conducive to moral progress.

The extension of new scientific knowledge is influencing morality and gradually reducing the selfishness of mankind, by proving that their exist no royal roads to happiness, and that the greatest amount of individual and national success can only be secured by a genuine pursuit of truth, as an individual and cosmopolitan duty. Increased knowledge is gradually proving to mankind that the purest happiness is to be obtained by intelligent and virtuous conduct. By shewing Man the unreasonable character of some of his fears and hopes, and substituting for them a greater variety and extent of intellectual pleasures, science is slowly making him more satisfied with his lot on this Earth. Meanwhile the great mass of mankind are still pursuing the ever retreating phantom of an easy way to happiness; the great laws of nature however cannot be evaded, the avoidance of evil and the attainment of good can only be secured by obeying all the great laws which govern our nature.

Progress in morality is largely dependent upon the diffusion of belief in the universality of scientific laws. When men understand those laws, know that their action is irresistible, and that they have no alternative but to obey them or suffer, they acquire a habit of obeying them. Universality of law in moral actions is often considered to be incompatible with the existence of freedom of the will in selecting ideas, and choosing courses of conduct; but we are free or not, according to circumstances, both to think and to act. All things are free to be active or not, in accordance with their properties and surrounding conditions, but not in contradiction to them ; and the human will is no exception to this statement. The " will " is only free within certain limits ; it cannot act in opposition to its strongest motives or causes of action. We believe ourselves to be much more free than we are, because we often do not know the causes which determine us, and we frequently fail to detect those influences, because we cannot think, and at the same time clearly observe our act of thought and its motives. Freedom of the will does not enable us to set aside laws : entire freedom from law in any instance is probably only apparent, This limited degree of freedom of the will indicates the dependence of volition upon scientific laws, because a supernatural power, being entirely independent of natural law, could not be limited by it. To affirm without proof that the human will is a " supernatural power " is to implicitly deny the universality and constancy of natural laws. New knowledge developed by Science, imparts to us liberty, but not license ; and, so far from diminishing the freedom of the will, increases it by showing us what conditions we must fulfil and obey in order to effect our objects. We acquire power by being first obedient ; and this is in accordance with the principles and facts of science ; we must obey nature before we can make nature obey us ; the elementary bodies, also, usually acquire the free state, latent power, and the ability to evolve heat and electric energy, only by being first subjected in their crude state to a process of reduction and purification.

Few circumstances connected with the discovery of new truths of science, have had a greater moral effect, than the very high degree of certainty of such truths. The moral result of this is a corresponding degree of confidence in the statements of science. Trustworthiness is a great moral quality. Uncertainty is a continual hindrance to action and enjoyment; and many persons are driven to believe in error, and hence to commit sin, rather than remain in suspense. Contradictions of doctrine, and the consequent uncertainty of belief, in any subject, are fertile sources of strife. Science consists, not merely of opinions and words, but also of the tangible realities which those opinions and words represent, the forces, substances and phenomena of the material Universe. Some persons however fancy that the results of science are as uncertain as those of the undemonstrable subjects with which they are familiar.

Another way in which science has contributed to moral progress, has been by requiring greater accuracy in nearly all human actions, and thereby diffusing greater exactitude of language and of conduct, which has spread itself throughout all civilised society. Previous to the use of watches and clocks, persons were no doubt much less exact in fulfilling their appointments; the establishment also in our chief towns, of electric time-keepers regulated from Greenwich Observatory, is increasing exactitude in our large communities. Since the introduction of railways, millions of persons have been compelled to be more exact in their movements, by the risk they incur of missing their train. Numerous inventions and processes based upon scientific discoveries could not be worked at all unless men possessed habits of greater accuracy than formerly. Workmen now require higher moral and intellectual education, and their duties require more intelligence and involve greater responsibility.

Science diminishes error, and the avoidance of error is a large step towards the attainment of truth. There is no tyranny equal to that of false ideas. Error often produces immoral acts, and every act of immorality is a mental error. " The ignorant justice-loving man, enamoured of the right, is blinded by the speciousness of wrong." " Inaccuracy of thought is the cause not only of the errors we meet with in the sciences, but also of the majority of the offences which are committed in civil life,—of unjust quarrels, unfounded law suits, rash counsel, and ill-arranged undertakings. There are few of those which have not their origin in some error, and in some fault of judgment, so that there is no defect which it more concerns us to correct."*

Our senses and consciousness are often great deceivers, and unless corrected by sufficient knowledge, are frequently as great a source of error in moral questions as in mental ones.† Their incessant influence is a cause of selfishness, and of the fallacious tendency existing in nearly every man, to exaggerate the importance of himself and of everything relating to him.

* Port Royal Logic, Discourse 1.   † See p. 91—92.

It has led man to consider himself "the Lord of Creation";—to believe that his volitions are not subject to natural laws &c. :—and has given rise to the comparatively narrow-minded idea "the study of mankind is man." The fact that consciousness frequently misleads men of energetic temperament who feel their energy, indicates its connection with a physical basis.

Consciousness is also an essential condition of what we term evil. If we define evil as that which produces pain or discomfort in sentient creatures, then evil is that influence only which unpleasantly affects consciousness. And if we admit this, then all evil is relative, and there is no absolute evil; because, if there were no sentient creatures, there would be no evil. It is manifest also, that if the existence of evil is dependent upon that of sentient creatures, and if the existence of such creatures depends upon physical conditions, and upon the operations of the great principles of science, then the existence of evil must itself depend to that extent upon those conditions and principles. What we term Evil, is caused, not only by the actions of man, but also on a large scale by the operation of the simplest forces of matter, in earth-quakes, storms, volcanic outbursts, droughts and famines, pestilences, etc. Evil (as well as good) may therefore be viewed as a result, to some extent, of the operation of the laws of the Universe; and here again we are compelled to recognise a scientific basis of morality.

That the same causes, acting under different conditions, produce different and even opposite effects, is a well-known scientific truth. The same heat of summer which causes our foods to decay, promotes the growth of plants in the soil: the same cold of winter which increases the pain of bronchial affections, and cuts short the lives of aged and infirm persons, acts as a stimulant and a source of pleasure to the young and healthy. We need not therefore be surprised, that the same physical conditions and principles of nature, act as causes or conditions both of what we term evil and what we term good. If, also the theory of relativity in physical and mental action is true, that change of impression is a necessary cause or condition of consciousness, and that previous experience of pain increases the perception of pleasure, we possess in that theory, as one of the general ideas of science, a partial basis of morality. All these remarks tend to shew, that in order to obtain a truly scientific view of the nature of man, and of man's position and duties in the Universe, we must avoid the errors caused by uncorrected consciousness.

Another great moral effect of the continual discovery of new truth in science, is the gradual production and diffusion of uniformity of belief, first amongst scientific men, and then amongst the mass of mankind. This uniformity of belief is a necessary result of the invariability of fact and law; it does not extend to scientific opinions, hypotheses or theories, because they are not necessarily facts, and may be

erroneous. A knowledge of science tends to remove
differences of opinion between man and man, because
it enables every honest examiner to obtain essentially
similar results. Scientific research will gradually disclose what is true and what is untrue in doctrine and
empirical rules ; and what is true will be retained
A universal religion or a scientific philosophy which
is composed of contradictory creeds cannot be wholly
true. Science is gradually superseding unreasonable
beliefs, and inaugurating a true universal gospel in
which all men will eventually think alike in fundamental matters. The continued discovery of new
truth must of necessity sooner or later lead mankind
to the source of all truth and to universal satisfaction
and happiness. It has been frequently stated that
science is antagonistic to religion ; it is evident
however that as science is so conducive to morality, it
cannot be opposed to true religion, but only to false
or unfounded beliefs. Nothing shews more plainly a
weakness of moral confidence and a deficiency of
faith in an over-ruling power, than a fear that the
pursuit of scientific truth will lead to results injurious
to mankind. What we most need to fear is, not that
our most cherished doctrinal beliefs may be proved to
be mistakes, but that we through deficiency of knowledge may be led to do wrong.

There are plenty of questions, especially in matters
of theory and doctrine in concrete subjects, which
science cannot directly and absolutely decide either
one way or the other, but respecting which, by the

aid of new knowledge and of inference based upon it, science gradually accumulates so large a preponderence of evidence as conclusively settles them to the conviction of every unprejudiced and reasonable person. Many of the most deeply interesting questions in mental science and morality are of this kind ; and will probably be settled in this manner. It is well-known also to scientific men that the indirect conclusions of the intellect and reasoning power are often more certain than the direct evidence of the senses and consciousness ; we are more certain for instance that the Earth is a sphere than that it is a plane, although the former conclusion is arrived at largely by inference, whilst the latter is the direct testimony of uneducated consciousness. Whilst our senses and consciousness inform us that the Earth is a fixed body, inference proves to us that it is rushing through space at an immense velocity. Sense and consciousness are not intellect, although they are often treated as such. Their functions are to perceive and observe, to act as witnesses, to supply evidence to the judgment, and not to usurp the reasoning power Even the universal consciousness of all mankind is insufficient to overthrow the final decisions of the intellect, or to decide what is true or false, because the senses and consciousness cannot compare or infer. As it is the force and repitition, and not the truthfulness of mental impressions, which largely determines belief, we are capable of believing error as well as truth, and we believe much that is erroneous until the corrections of the intellect are applied to the evidence of the senses and feellngs. The correctness or otherwise of our present beliefs will be tested in the future as others have been in the past, and the new experiences requisite for the purpose will probably be obtained by means of original research. It is a great mistake to suppose that the warrantable inferences deduced from scientific knowledge will not sooner or later profoundly influence questions relating to the highest hopes and aspirations of the human race, such as the independent existence and immortality of the human soul; that of a personal Ruler of the Universe; freedom of the will; the origin of evil; future reward and punishment; &c. By extension of knowledge a scientific system of morality will be formed. The great principles which govern the phenomena of all bodies are gradually being discovered, and when found we deductively apply them to ourselves, and thus arrive at a knowledge of our true position in nature, our duties, our proper course of conduct, &c. Science also by disclosing to us the true relations of matter to mind in the human brain, will probably not only make known to us the true limits of our mental powers and of the knowable, but also help to solve the problems of the relations of the Universe and of Man to an intelligent Creator. It will decide such questions, largely by shewing us whether or not the ideas we entertain respecting them are consistent with the more extensive knowledge evolved by research. A part of the data from which we may safely predict that science will in the future exercise so great a moral influence over mankind, is the fact that its chief principles are fundamental guides and regulators of human action.

Probably nothing has a greater effect in making a man humble and reverent than a thorough knowledge of science.   By the inventions of the telescope, micro-scope, spectroscope, telegraph, microphone, telephone, &c., the extremely finite extent of all our faculties has been abundantly demonstrated.   Whilst the wonders of the telescope have developed an intelligent sentiment of reverence, by revealing to us a portion of the vast amount of the Universe of matter and energy, those of the microscope have strengthened that sentiment by affording us an insight into the almost endless com-plexity of minute creatures, substances and actions. Whilst also these and other scientific instruments and appliances have proved the excessively limited ex-tent of our senses ; the inscrutable character and immense number and variety of problems of nature yet unsolved, equally demonstrate the extreme feebleness of our mental powers.   To obtain an ac-curate acquaintance with science also, and especially to discover new scientific truths, it is absolutely necessary to set aside human pride, and approach the subject like a little child ; no other course is possible.

A knowledge of geology and astronomy also makes a man humble and reverent.   The fact that this globe must have existed myriads of years ; and is always moving at the immense velocity of more than 62,000 miles an hour in its orbit, is sufficient to convince any unprejudiced person of his own transient physical existence and his comparative physical feebleness and insignificance.   Hitherto, man has largely been accustomed, through the influence of uncorrected impressions and other causes, to view all nature as having been expressly provided for him, but science informs us that whilst this Earth is suitable for his abode, and Nature ministers to his necessities and pleasures, it is only on condition that he first obeys the great laws of matter and energy, and adapts himself to their requirements. The operation of those laws often ruthlessly destroys thousands of men by pestilence, famine, drought, and other great calamities, and man can do nothing which is incompatible with them without suffering a penalty. Science shews that man is but one out of at least 320,000 different species of animals; it also discloses the fact that the entire human population of this globe constitute only about one 50,575,785 millionth part of the Earth, and proves to us that the Earth itself is but a speck in the Universe, one out of at least 75 millions of worlds ; and that not only is it merely a planet revolving round the Sun, but that the Sun is only one of a multitude of Suns, and is itself, with all its planets, revolving round a still more distant centre in space.

There is scarcely a faculty man possesses, which is not immeasurably limited in comparison with the powers and capabilities of inanimate nature. His physical energy, when compared with that of the momentum of this Earth, is so exceedingly small that it can hardly be conveyed to our minds by means of figures ; even the steam engine, excessively wasteful as it is of power, far surpasses him in strength. The duration of his existence is to that of the world he inhabits, as nothing to infinity. His power and speed of locomotion are also very limited ; the globe to which he is fixed by gravity, moves in one hour through a distance greater than he could walk in twenty years. Practically, by circumstances, he is almost rooted like a vegetable to the locality where he exists ; comparatively few men have walked even a hundred miles from their homes, or have been conveyed round this little globe by the aid of all our improved means of transport. A balloon can ascend in the air, but a man cannot ; without the aid of that apparatus he is absolutely fixed to the surface of the Earth, and with the assistance of all the appliances of science, he cannot yet ascend even ten miles into the atmosphere, nor dive more than a few fathoms into the sea. His senses are equally contractcd ; his perceptions of touch and sound are far less delicate than that of the microphone ; a photographic surface will detect vibrations of light which he cannot at all perceive, and record images more quickly than his brain ; and for the detection of magnetism and the chemical rays of light he possesses no sense whatever :—electrometers and galvanometers can detect thousands of times smaller quantities of electricity than he can perceive :—whilst a bolometer renders manifest a one hundred-thousandth of a Centigrade degree change of temperature, he can hardly detect a difference of an entire degree; and whilst carbon and platinum may be heated to whiteness without material change, a rise or fall of about five Fahrenheit degrees in his temperature endangers his life. His mental and intellectual powers are as limited as his senses ; he can hardly reckon without making an error even a single million, nor can he conceive an adequate idea of a billion ; a million miles or a millionth of an inch are each quite beyond his immediate perception ; an extremely minute circumstance also is capable of disturbing and entirely diverting his train of thought. He cannot create or destroy even a particle of dust, nor form out of nothing a single idea. The velocity of transmission of his nervous power, and the speed of his execution of will, are also extremely slow in comparison with that of an electric current in a copper wire. Every person is aware that he can only very slowly receive and understand a new idea. His mental advance is as tardy as his locomotion, a sixth part of his life is spent in acquiring the merest rudiments of universal knowledge. Whilst his reasoning power, when applied to actual and truthfully stated experience, is truly "the great guide" of his life, it only renders explicit what was already contained in that experience ; for when he draws an inference, he usually only states in one form of words, what he has already implicitly included in the propositions ; and if the inference contains more than this it is unwar-

ranted.   His mental helplessness in the absence of
knowledge, is equal to his physical incapacity in the
absence of light.   Nearly every problem of nature
also is so complex, and affected by so many conditions, that his reasoning power only enables him to
advance a very minute step at a time in the discovery
of new knowledge ; he is then obliged to halt, and
have recourse to new experiences obtained either by
means of experiment and observation, or by the latter
alone.

Man's moral actions are largely the effect of
circumstances ; his thoughts and actions are probably
the whole of them limited by law.   He is never free
from the influence of causation.   His mental and
moral freedom are limited by the epoch in which he
lives, by the customs of his nation, by the individuals
by whom he is immediately surrounded, by the
alcoholic stimulants of which he partakes, and by his
own physical and mental constitution, his degree of
intelligence, &c., &c.   Whether he is willing or not,
he is incessantly compelled to receive sensuous and
mental impressions, and be influenced by an almost infinite number and variety of agencies acting upon him
both from within and without :—To be mentally and
physically active, and perform all the bodily functions
and acts necessary to his existence :—To live on this
globe in presence of all its phenomena, and be carried
through space at an immense velocity :—To undergo
through a long series of generations a progressive existence and development of civilization, &c., &c.

He is more subject to the laws of the Universe than those laws are subject to him ; and he can only exercise his will successfully and become their master by first obeying them.

Under the influence of the light and heat of the Sun, the entire population of this planet (about fifteen hundred millions) are renewed out of the crust of the Earth every few years, by breathing the air, drinking the water, feeding upon plants which take their constituents from the Earth, water and air ; or by eating animals which have lived upon plants ; and if that heat and light, or that supply of food and air, were to cease, all those human beings would die, and all the moral phenomena of man on this globe would terminate. Whilst man cannot exist without the support of inanimate nature and the operation of its laws, inanimate nature and its laws can exist without him That also which is naturally ordained by Creative power to be dependent, cannot be essentially more important than that upon which it depends for its existence. The essential importance of man in relation to the Universe, exists only in his own imagination.

These facts shew that the principles of science, and the physical and chemical properties of substances, lie at the very basis of man's existence and activity and it would therefore be incorrect to say that the physical system of the Universe is unimportant in comparison with the moral phenomena of mankind.

That science conduces to humanity by preventing

and alleviating animal suffering has been already alluded to (p. 80—81). True humanity consists not in the abolition of experiments upon living creatures, but in the judicious employment of them. Instead of barbarously treating our suffering fellow creatures by indolently and ignorantly allowing causes of disease and pain to continually occur and take their course, it urgently enforces upon us the duty of extending our knowledge of physiology by means of new experiments, observations and study. It would be untruthful to say that experiments purposely made upon men and other animals do not yield new and valuable information;—Pharmacopœias and Materia-Medicæs are full of descriptions of the properties of curative agents discovered by these and other scientific methods.

Amongst the lesser virtues which have been greatly promoted by means of scientific research is that of cleanliness. The origin of soap was the discovery of the detergent properties of a boiled mixture of fat and alkali. The numerous inventions which have cheapened the most important soap-producing material, viz., washing soda, and those which have cheapened oil of vitriol, the chief substance consumed in making washing-soda, have all contributed to the cleanliness of mankind ; and it has been stated that the degree of civilization of a nation might be ascertained by the amounts consumed of those substances.

Even the minor virtue of economy has been greatly promoted by the results of scientific research. New scientific truth has through inventions taught us how to obtain greater effects with less expenditure of space, of time, of materials, and forces. It has enabled us to effect our objects quicker and with a dimunition of waste. In the sugar manufacture for example, by means of the centrifugal machine, the sugar is deprived as perfectly of molasses in three minutes, as it was previously in three days, and the necessary manufacturing apparatus has been so much reduced in magnitude as not to require more than one half the space. The process of bleaching linen, which formerly required weeks, has by the discovery of chlorine been reduced to hours. Journeys which at one time occupied weeks now only require days. Messages are now transmitted in hours which formerly required months. Multitudes of instances might be adduced of the diminished cost of the comforts and conveniences of life, resulting in consequence of dis- covery of new scientific knowledge. Ultramarine for example, which at one time cost from ten to twenty pounds an ounce, has by means of chemical research been reduced in price to a few pence per pound ; phosphorus, which formerly cost several guineas an ounce, now costs only as many pence.

Numerous substances which were formerly thrown away, destroyed, or neglected, are now utilized. Coal tar and gas-water, which were at one time waste products in the making of gas, and which when thrown away were the causes of costly litigation to gas-companies, by polluting streams and wells, &c.,

are now sources of very large income to those companies. Those substances yield great quantities of salts of ammonia, the beautiful aniline dyes, paraffin, benzene, napthaline, alizarine, and other valuable products, Glycerine also, which formerly was a most offensive waste product in soap-making, is now purified and used, to an extent of twenty millions of pounds anually, for a great number of purposes; as an emollient for the skin; as a source of nitro-glycerine and dynamite, used in blasting rocks, in warfare, &c. The immense beds of native sulphide of iron also, notably those of Tharsis and Rio Tinto in Spain, and of many other places, are now utilized, literally in millions of tons, for the production of sulphur, copper, oxide of iron, &c. A long list of instances of this class might be adduced if it were necessary, some of them of very great importance.*

The promotion of morality by enabling us to detect crime, is one of the smaller influences of scientific research, and may be referred to as a set-off against the bad uses sometimes made of scientific knowledge. The telegraph is very commonly employed to assist in tracking and capturing criminals. Photography is also largely used in our goals as a means of recognising offenders.

Knowledge of science conduces also to self-discipline and self-mastery, it tends to bridle our vicious passions by making known to us the penalties which must be paid for their indulgence; it limits our self-will by shewing us that we must respect and obey the

* See " Waste Products and Undeveloped Substances," by P. W. Simmonds.

laws of nature whether we are willing or not, no man can improperly manipulate dangerous substances or forces with impunity ; it moderates our bigotry by exhibiting to us the great uncertainty of unproved opinions ; it restrains undue credulity in men's assertions, by shewing us their frequent fallacy ; it gives us confidence in the laws of nature, by proving to us their uniformity ; it withdraws us from self-deception by compelling us to accept the truths of nature as they exist ready made for us, whether they harmonise with our preconceived ideas or not ; men cannot argue with nature, as they can with their fellow-men, but must submit to the influence of verified truth. It supplies us with principles instead of empirical "rule of thumb" methods as guides of morality. Whilst it liberates us from the terror of irrational fears, it cautions us against entertaining unreasonable hopes. It substitutes for ignorant wonder and awe, an intelligent appreciation of created things; and when fully developed it will probably satisfy all the reasonable instincts and desires of men.

Whilst law, medicine and divinity, direct man's attention almost exclusively to matters concerning himself, and thus tend to limit his sphere of perception and knowledge, and unconsciously impress him with the idea that all other existences are less important than himself, science not only enlightens him respecting all the departments of his own nature, but extends his mental vision in all directions by exciting his mind to observe and reflect upon all other bodies and actions throughout the Universe. Whilst also music, painting, sculpture, poetry and the drama, afford excitement and pleasure to his senses, feelings and sentiments, and are largely personal ; science not only constitutes the basis of those arts, but shews the relations of them to Man and to the external Universe, and thus more largely cultivates the intellect and corrects and refines the senses, feelings and sentiments.

New scientific knowledge affords advantages to all classes of men ; to the minister of religion, by supplying him with new illustrations of Creative power, in the greatness, smallness, and vast variety of nature ; to the physician, by explaining to him more perfectly the structure and phenomena of the human body, and by providing him with new remedies ; to the statesman and politician, by making known to him the great and increasing relations of science to national progress, by its influence upon wages, capital, the employment of workmen, the art of war, the means of communication with foreign countries, &c. ; to the philanthropist, as an endless source of employment for poor persons, by the development of new discoveries, inventions, and improvements in arts and manufactories ; to the military man, by affording him new engines and materials for warfare and defence ; to the inventor, by supplying him with new discoveries upon which to found inventions ; to the merchant and man of trade, by the influence of new products and processes upon the prices of his commodities ; to the manufacturer, as a means of improving his materials, apparatus, and processes ; and to the investor of money, by assisting him to judge what new technical schemes are likely to succeed.

As the domain of rational enjoyment afforded by means of science gradually enlarges, that derivable from less intellectual sources will probably be modified ; indeed this change is already progressing, and is manifested in the alterations occurring in theological views, and in the extensive adoption of scientific entertainments by religious bodies. The recognition of science by professors of religion is also shewn by the already extensive use of railways on Sundays as a means of conveyance to churches and chapels ; also by the publication by the Society for Promoting Christian Knowledge, of Manuals of Electricity, Astronomy, Botany, Chemistry, Crystallography, Geology, Physiology, Zoology, Matter and Motion, the Spectroscope, &c.

Having shewn some of the chief modes in which new scientific truth is a basis of mental and moral progress, it is not necessary to say much respecting the evil uses sometimes made of science, because every good thing is liable to be abused by ignorant or ill-intentioned persons. The abuses of scientific knowledge do not arise from the true spirit of research, viz., a desire for new knowledge on account of its intrinsic goodness and value to man, but from an absence of that sentiment. The Bremerhaven explosion, the assassination of the Czar, the uses of photography to forge letters of credit, and of the telegraph in swindling operations, the employment of electro-gilding and silvering in coining base money, &c., &c., are all attributable to motives other than a love of science.

All the facts mentioned in this chapter, and the various points of essential similarity between physical, physiological, and mental phenomena, justify the conclusion that both moral and other mental actions, like physical and chemical ones, are obedient to the great principles of science. And from the evidence here adduced and alluded to, it is certain that those principles influence human progress, not only in a few conspicuous direct ways, but in a multitude of varied, deep-seated, and indirect ones.

If the statements made in this Chapter are true, that the innate properties of matter really are motive powers of the human organism, and the principles of science are regulators of mental and moral action ; that Man is a feeble epitome of the principles and powers of inorganic matter ; that the laws of Nature operate in utter disregard of his erroneous beliefs ; that nearly all man's sins and sufferings are traceable to his ignorance and limited powers ; that in proportion to his ignorance of science so is he unable to foresee the more remote consequences of his thoughts and acts ; and if new knowledge does correct erroneous beliefs and purify human thought and action, it behoves teachers of morality to make themselves adequately acquainted with the principles and newest developments of science.

CHAPTER III.

New Truth, and its General Relation to Human Progress.

The great source of the success of applying science to trade, and of the beneficent effect of science upon human welfare in general, is simply the influence of demonstrable truth. We know that if we have once discovered all the principles, laws, and conditions of some scientific phenomenon, or of some improved process or result in a manufacture, the reproduction of exactly the same conditions will hereafter enable us to invariably produce the same result. In this respect science differs from dogma, the truth or falsity of which cannot be demonstrated ; it also differs from empiricism, because when empirically working a process we are ignorant of the principles or laws which are operating, whilst with a scientific knowledge we understand those laws, and can direct them to our particular purposes. In the process of electro-plating for example, we understand the laws of the phenomena, and can direct them so as to obtain silver of a hard or soft quality, brittle or tough, crystalline silver, &c., according to our wish ; but if we had only an empirical knowledge of the subject we could not thus vary the process.

The highest test of truth is verified prediction ; if we calculate beforehand that an eclipse of the Sun will occur at a certain hour and minute, and that eclipse occurs accurately at the predicted moment, we may rest assured that our knowledge upon that point is true and complete. If we say that a piece of clean iron, immersed in a solution of blue vitriol, will become covered with a layer of metallic copper, and we find upon trial that this result invariably occurs when we fulfil those conditions, we may be certain that our knowledge of this phenomenon and its conditions is also of a definite and certain character. Similarly, when we become able to predict with certainty the conditions of the Sun's surface, we shall probably also be able to predict severe winters, famines, &c., and therefore be prepared to suggest precautions to be taken against them. Even now the new truth necessary for this purpose is beginning to be evolved by means of scientific research.*

Amongst the great axioms and principles of science, possessing great certainty, and which enable us to predict, are, *1st.* the general truth known as the Principle of Causation, that every effect has a cause; that the same cause, acting under the same conitions, always produces the same effect ; and that causation acts through all time and all space :—*2nd.* the great truth, that every phenomenon requiries time ; and every substance occupies space :—*3rd.* the Principles of Conservation and Persistency of Matter

---

* See " Barometer Cycles," by Balfour Stewart, F.R.S.—*Nature*, Jan. 13. 1881, p. 237.

and Energy; that out of nothing, nothing comes; and out of everything, everything proceeds; that all the future states of the Universe are implicitly contained in and will be evolved out of the present state of the Universe; that we have no experience and possess no verified knowledge either of creation or annihilation of Matter or Energy; that we cannot absolutely create or destroy even an idea; * and that Matter and Energy appear to be eternal :—4*th*. the Principle of Convertibility and Equivalency of the different forms of Energy, according to which the various forces known as mechanical power, heat, light, electricity, magnetism, chemical action, &c., being modes of motion, are convertible into each other in equivalent quantities and without addition or loss. These and other great principles constitute the basis of physical and chemical science, by obeying which, we have been enabled to evolve all the wonderful practical realities of science of the present day.†

To these great principles may be added the more concrete truth called the " Law of Progress," the essential idea of which is time, a time-rate; which regulates the speed of increase of civilization, and is evidently connected with the great truth that every phenomenon occupies time.

The Principle of Gravitation, demonstrated by Newton, explains a vast number of facts relating to the motions of the Heavenly bodies :—the Undu-

* See p. 165, et seq.

† It would I consider be an improvement in our educational arrangements, if a Professorial chair, solely devoted to teaching those laws and principles, existed in each Scientific College.

latory Theory of Light, largely developed by the
labours of Fresnel, renders equally clear and syste-
matic an almost endless number and variety of
optical phenomena; Oersted's law of Electro-magnetism similarly explains and renders consistent a
multitude of facts respecting the movements of
magnets and electric conductors, which would otherwise be confusing to remember and impossible to
satisfactorily explain. And the great mental value
of these comprehensive ideas to mankind, consists
largely in relieving the memory and diminishing
mental confusion, by co-ordinating a large number of
different facts and apparently inconsistent phenomena by means of a general conception which embraces
the whole of them. Thus a knowledge of the Principle of Gravitation informs us that both the ascent
of a balloon in the air, and the descent of a stone in
water, are alike due to the same force of gravity ; and
that of Chemical Affinity proves to us that the apparently unlike phenomena of slow rusting of iron and
vivid combustion of phosphorus are essentially alike
and due to the same cause.

All bodies, whether living or dead, and all forms
of energy, appear to be absolutely subject to the great
laws of Causation, Progress, Conservation, &c., no one
can escape them ; the man who transgresses the Law
of Progress by being too much in advance of his
epoch, is punished as certainly as he who lags behind
it ; all must advance together, and at approximately
the prescribed rates.

The real source of all that is good in new scientific knowledge arises from its verified and verifiable character, its high degree of certainty, and its capacity of withstanding all the tests which can be applied to it. By the term "scientific knowledge" in this case I mean that only that which has been verified, and I purposely exclude all matters of hypothesis, mere opinion or belief. Scientific research is the chief basis of national progress, not only because it is continually disclosing new truths to us, but also because the truths it reveals are frequently of the most definite kind.

As the term "verified truth" may appear vague, the questions may well be asked, what is truth? And how may we best detect it? And especially what is new truth? and how may it best be recognised? Truth may be conveniently defined as universal consistency; or that which perfectly conforms to facts, and agrees with the widest experience, when tested by means of all our intellectual powers, the reasoning faculty in particular. The usual modern criterion of it, is consistency with the fundamental axioms of logic, and with all the great principles of nature as established by means of scientific research, such as the universality of causation, the continuity of phenomena, the indestructibility of matter and energy, the convertibility and equivalency of forces, &c. All truth whatever is one in character by possessing the inseparable attribute of complete consistency. The truthfulness of scientific knowledge is proved by its agreement with universal experience and with the fundamental logical axioms :—a thing either is or is not :—-a thing cannot both be and not be :—a thing must either be or not be :—things equal to the same are equal to each other ; &c. It is chiefly by means of knowledge of these axioms and of the above principles of science, and of their varied and numerous modes of operation and application, that the man of science "explains" the multitudinous phenomena of nature, predicts future events, and is enabled to discover new truths and develope new inventions in the arts. Unlike other persons ; when he sees a new effect, or hears of a new phenomenon, he at once refers it to these principles, in order to test its correctness or to explain it.

With regard to the detection of truth, that is often a difficult and complex process. There exists no royal or easy method; usually it can only be recognised by means of laborious and critical examination of the whole of the evidence obtainable in the case ; and even then we are often obliged to be satisfied with only an approximation, or it may be with even a mere probability. Frequently also the truthfulness or otherwise of a statement cannot be decided in any degree in consequence of the absence of suitable or sufficient evidence, and for that we may have to wait for ages. We are now waiting for evidence necessary to decide many questions respecting the human mind.

With regard to the question, what is new truth ? ; that also is a difficult one to solve. The forms in which different truths appear are so various, and those also in which even the same truth may shew itself are so diverse, that it is often impossible to discriminate new truth from old ideas clothed in a new form of words. The newness of an idea is entirely a question of evidence, and to determine it, usually requires a complete knowledge of all the circumstances affecting the particular case.

New truth appears to be usually derived from new physical or mental experiences of phenomena external to our perceiving faculty ; either by observing matter or its forces under new conditions or from a new aspect; and the knowledge comes to us either through the avenues of our feelings and senses, or by means of direct observations, by comparison of such impressions, or by inferences drawn from them. From the results of such mental operations, additional new truths are evolved by the more complex process of analysis, combination and permutation of ideas. New truths are also evolved from old ones by each of these latter methods ; but sooner or later the implicit contents of our stock of old knowledge becomes exhausted when used for such a purpose, and we are then obliged to seek new experience.

As new truths may be acquired in the more direct manner, by acquisition of new experience ; and less directly, by mental operations upon old ideas, other subjects of less fundamental and more concrete nature than the simple sciences, such as sociology, &c., are also sources of progress, when treated in these ways.

Of all subjects, the simple sciences of physics and chemistry, are at the present time, apparently making the most rapid advance, and the chief reasons for this probably are, 1*st.* they treat of facts and principles which can be verified, and 2*nd.* because the more complex sciences, together with the arts and manufactures based upon them, can only improve in proportion as they are developed. All the essentially human subjects, such as sociology, politics, morality, religious worship, &c., are in this position, and are probably results partly of the operation of the great principles of nature acting through the body and mind of man.

The chief method of discovering new truth is that of observation, experiment, and study, and further mental treatment of the results. The most systematic methods also of evolving new truths are those employed by scientific men in making discoveries, and when any person arrives at a new idea, he usually (either consciously or unconsciously) employs them.

The acquisition of new knowledge must of necessity precede its diffusion. Immediately a new truth, especially an important one, is discovered, its influence begins to permeate the existing mass of knowledge in various directions, causing us to view many of our old ideas in a new aspect ; giving rise also by comparison and inference, and by processes of combination, permutation and analysis of ideas, to a multitude of other new truths, usually less important ones, which themselves also affect previous knowledge in similar ways, and by analogous treatment give rise to additional new conceptions.

But although we evolve truthful new conceptions from previous ones by these purely mental methods, there is a limit to the number capable of being evolved from a limited stock of ideas, because the number of combinations and permutations of such a stock, though usually large, are themselves limited. The number however is large in proportion to the degree of essential importance of the ideas, and is greatest when we employ those of the fundamental principles of nature, already referred to ; for instance a greater number of new ideas have been evolved by means of appropriate mental processes from the law of gravitation and from that of electro-magnetism than from any minor truth in science. Persons therefore who are the least familiar with great demonstrable principles are usually the least able to conceive new truthful ideas of intrinsic importance, or to draw new verifiable inferences of much theoretical value.

Every inventor and student knows that he continually requires new materials of thought, and if he does not obtain them, an obstacle, like a wall of adamant, rises before his mind in all directions, and prevents his forming new ideas. That also which is true of each individual is true of the collection of individuals, mankind; if new truths are not obtained, the thoughts of men flow in circles, and mental progress ceases. The mental characteristics of sequestered communities in remote isolated districts, are examples of this fact. The influence of printing, railways, telegraphs, postal communication and other scientific developments, in aiding mental progress, afford other illustrations. A multitude of facts of this kind, and many others, leads us to the conclusion that each new idea requires a cause to produce it, and that human knowledge is subject to the great law of causation; also that the creation of an idea out of nothing would be a miracle, a phenomenon without a cause. Our present knowledge was not created by us, but was originated by previous knowledge and experience, including of course inherited impressions. Even in what is termed the "noblest effort of the mind," an act of reasoning or inference, we do not create an idea, but only render explicit in a new form of words, ideas already implicitly contained in the words of the propositions employed, as may easily be rendered manifest by mechanical means in Jevons's "Logical Machine;" a proper inference never contains more than its data. In the so-called "creation" of ideas by the imagination also, the new ideas are evolved from old ones, and rendered explicit by mental processes of analysis, combination, permutation, &c. Our scientific inventions also, being mental conceptions, an unlimited number of them cannot be made by means of a limited stock of old knowledge. It was in consequence of this limit, viz,, the impossibility of actually creating ideas out of nothing, that human knowledge was not more advanced by metaphysical speculations until science with its experiments and observations came to its assistance. These various facts prove the statement made in the Preface of this book, that present knowledge only enables man-kind to maintain its present state.

Not only the mental, but also the physical advance of mankind is essentially dependent upon the discovery of new truths. Men's physical actions are determined not alone by their inherited and acquired tendencies and the influence of external nature upon them, but also by their ideas ; as, a man thinks, so also to a large extent does he act. Nations who do not adopt new ideas do not either mentally or physically advance, but change only so far as their immediate surroundings change ; the Chinese are a remarkable example of this ; even the tendencies which men in-herit, were largely produced in their ancestors by the influence of ideas. The great fact of the essential dependence of human progress upon new knowledge, is a truth, the importance of which to man cannot be over-estimated, and is one which statesmen, ministers of religion, and philanthropists should seriously study.

Much of the apparent advance of this nation how-ever is not real. The great bulk of our newly published knowledge, even that which is scientific, and considered by the public to be new, consists, not of new truths, but of old ones dressed in new forms of expression :—

> " The tale repeated o'er and o'er,
> With change of place and change of name.
> Disguised, transformed, and yet the same
> We've heard a hundred times before."—*Longfellow.*

It falls to the lot of but comparatively few men to discover or evolve important new truths. The great majority of learned men also, have through all historic time been occupied, and are still, not in evolving new ideas, but in re-expressing old ones in different forms of words; the literary spirit, is in civilised nations, almost universal. In ordinary writings it is a rare circum‾ stance to meet with an important and really new idea; it is usually in books written by men who are acquainted or imbued with the great principles and truths of science, that the newest demonstrable ideas are most frequently found. The difference between evolving new ideas, and re-expressing and permutating old ones; largely characterises the dissimilarity of the scientific and the literary and theological minds. All however are necessary to the welfare of mankind, the former to advance and the other to maintain the condition of man. If all were not necessary they would probably not exist.

There are two great artificial divisions of scientific knowledge also, upon the development of which national progress largely depends, viz., knowledge of inanimate matter and knowledge of man; the latter we have largely cultivated but the former we have greatly neglected :—and even our study of man has been largely one-sided and literary. It is far less important to know what is man, than to know what are the great principles which underlie the actions of all living creatures, and in obedience to which man is compelled to work out his destiny in the Universe and the infinite future.

It is evident from this, that much of the mental activity around us is not progress, but rather a process of maintaining present state, a prevention of decline, a continually going round and round in conventional varied step, a kind of intellectual mill. Under these circumstances it is not the original discoverer, but he who in this occupation, can best express old ideas, in the most varied forms and choicest language, who is most generally considered to be the greatest intellectual chief.

Although originality even in literature and art is very imperfectly encouraged in this country, both art and literature are much more readily understood and appreciated than scientific research, and treated as if they were more important. Whilst most persons can understand and appreciate the gift of a work of art to a public art collection, few can understand or properly value the discovery and gift of a new scientific truth of far greater intrinsic value to the public stock of knowledge ; the treatment received by Priestley and other discoverers in comparison with that of local donors, sufficiently illustrates this. Even publishers of lucrative newspapers prefer to give prizes and pay liberally for sensational tales, than to pay for articles on the public advantages of new scientific knowledge.

CHAPTER IV.

THE PROMOTION OF SCIENTIFIC RESEARCH.

NEARLY the whole of the most distinguished mathematicians, physicists, chemists, biologists, and physiologists of Great Britain, also the Earl of Derby, the Marquis of Salisbury, Sir Stafford Northcote' and many other eminent men, have given evidence before the Royal Commissioners on "Scientific Instruction and the Advancement of Science"* to the following effect :—1st. That the promotion of original scientific research is neglected in this country. 2nd. That such research is encouraged more by the State in Germany than here. And 3rd. That much greater encouragement of it by our Government, by the Universities and the Public, is highly necessary to our commercial prosperity. They have also stated in evidence their opinions as to the best ways in which they consider it may be assisted. The additional fact that all the greatest scientific men who have ever existed have pursued research, and sacrificed much for it, is a practical proof that they also approved of its encouragement.

That research should be promoted is further the opinion of many men learned in politics, literature, art, and science. The views expressed in numerous letters on the subject, received by me from Members

* See vols. 1 (1872) 2 (1874) of the Reports of that Commission

of the Privy Council, and of both Houses of Parliament, and from other eminent persons, confirm this. It has also been adopted as a chief part of the programme of the " Association for the Organization of Academical Study." *

" M. de Candolle, Corresponding Member of the Academy of Science, Paris, is a strong advocate for the encouragement of a class of sinecurists like the non-working Fellows of our Colleges, who should have leisure to investigate and not be pestered by the petty mechanical work of continued teaching and examining." " The modern ideas of democracy are adverse to places to which definite work is not attached, and from which definite results do not flow. This principle is a wise one for the mass of mankind ; but is utterly misplaced when applied to those who have the zeal for investigation, and who work best when left quite alone."

The correctness of the principle of promoting research is also recognised by our Governments in their yearly grants of money to the Royal Society, and to the Royal Irish Academy † to aid research, also by the Council of the Chemical Society, which has established a fund for the same purpose ; and by the British Association in their annual grants for the promotion of scientific enquiry. The Fishmonger's Company also presented to Mr. W. R. Parker, F.R.S., the sum of £50, followed by an annual gift of £20 for

* See pages 100 and 101.
† " *Nature*," April 3rd, 1873, p. 431.

the three years, to assist him in bearing the expenses of his researches on the skulls of vertebrate animals. And the British Pharmaceutical Conference voted the sum of £80 from the Bell and Hills fund, during the period of three years, in aid of research in connection with pharmaceutical science. A small fund for the purpose of research exists also at the Royal Institution. Fellowships also with a similar object have been founded at the Victoria University. Dr. Priestley also was aided in his researches by contributions from a small circle of friends. In recognition of the same principle, nearly all of the most eminent scientific men on the Continent have been assisted by their respective Governments. The total amount of aid to research in this country is however very small, and to one acquainted with the great commercial and other any valuable results of such labour, it is simply astounding that we have not systematically organized a powerful means of promoting discovery.

A few scientific persons however still continue to oppose aid to research ; quite recently, scientific investigators have been spoken of as a class of " men amusing themselves without any result whatever." * That idea   however abundantly refuted in the foregoing pages. It has also been remarked † that " practically, endowment of research comes to the creation of positions where there is payment without corresponding labour." " In England above all countries in the world, there will always be plenty of

---

* Sir Edmund Beckett, " *English Mechanic*", 1881, No. 830, p. 560.
† The Earl of Craufurd, " *Engl'sh Mechanic*," 1881, No. 830, p. 560.

amateurs ready and willing to assist in research, and it is notorious that in England, almost without exception, all the great advances in science have been made by such amateurs. Therefore I do not think it at all desirable that the British tax-payer should be required to put his hand in his pocket to provide salaries for gentlemen who might be working rightly or wrongly. He could not control them, and while there are such a body of amateurs in the country, I think the researches may be very well left to them."

The first of these statements is not correct; the endowment of research does not amount to "payment without corresponding labour." Scientific discoverers have always been distinguished as a body of men intensely devoted to their labours, and willing to perform much work for small payment. Most of the great advances in science also in England, have been made not by "amateurs," but by men of great experience, such as Newton, Herschel, Priestley, Davy, Faraday, Graham, and many others. Endowment of research is not desired for wealthy amateurs, but for investigators of proved ability and small pecuniary means and who require assistance. Such men, although not infallible, are the least likely to "work wrongly," and much less likely to do so than amateurs. Many scientific investigators also of repute, object to give their services wholly to a wealthy nation, because they cannot afford to do so, and because it is only just that the nation should make them some pecuniary return for their skill and labour. The great evils in this country requiring new knowledge to remedy them* also prove that there are not " plenty of amateurs ready and willing to assist in research here," or that " the researches may be very well left to them."

Whilst some investigators have had abundant means to carry on research, and have excelled in that occupation, many of the most eminent have been persons of limited circumstances ; and their insufficient pecuniary means has often restricted their degree of success. The argument also that insufficiency of means stimulates research, is only employed by persons who are not making investigations under such a condition.

The President of the Royal Society, Dr. Spottiswoode, in his recent address,† also remarked :—" The question has been raised whether it be wise, even in the interests of science, to encourage any one not yet of independent income, to interrupt the main business of his life. It is too often assumed that a profession or a business may be worked at half speed, or may be laid down and taken up again, whenever we like. But this is not so, and a profession temporarily, or even partially laid aside, may prove irrecoverable, and the temptation to diverge from the dull and laborious path of business may prove to have been a snare." Each of these remarks appears to be made upon the assumption that it is still a doubtful question whether persons qualified for research should be encouraged or not to abandon occupations they reluctantly follow, and for which they are less fitted, in order to become scientific

* See page 68, et seq. p. 134.
† See " *Nature,*" Dec. 2nd, 1880, p. 112.

discoverers. As it is a fact that the welfare of this country is suffering through deficiency of encouragement of research, it is certainly desirable to encourage, by every proper means, qualified persons to occupy themselves in such labour. Some of the greatest discoveries have been made by men "not yet of independent income," for instance, those made by Scheele, Priestley, Dalton, Faraday, W. Herschel, and many others.

The late Astronomer Royal, also,* who has made many researches, and was a scientific official paid by the State, says:—"I think that successful researches have in nearly every instance originated with private persons, or with persons whose positions were so nearly private that the investigators acted under private influence, without incurring the danger attending connection with the State. Certainly I do not consider a Government is justified in endeavouring to force, at public expense, investigations of undefined character, and, at best, of doubtful utility ; and I think it probable that any such attempt will lead to consequences disreputable to science. The very utmost, in my opinion, to which the State should be expected to contribute, is exhibited in the large grants intrusted to the Royal Society. The world, I think, is not unanimous in believing that they have been useful." He then enumerates what he considers "the proper foundations of claims upon the State," which he illustrates, and substantially includes in and

* "*English Mechanic,*" 1881, No. 831, pp. 586, 587.

limits by, the kinds of scientific research done under his direction at the Royal Observatory. He further adds—"The Royal Observatory was founded expressly for a definite utilitarian purpose (the promotion of navigation) necessarily connected with the highest science. And this utilitarian purpose has been steadily kept in view for two centuries, and is now followed with greater vigour than ever before. To its original plan have also been added—but still in the utilitarian sense—the publication of time, the broader observation of terrestrial magnetism, and local meteorology." His views therefore appear to be, that State aid to research should be limited to utilitarian objects; and that it is with propriety given to his own department, which is connected with the State. It has however been abundantly proved that nearly all the great scientific utilities of every-day life, had their origin in the pursuit, not of utilities, but of pure truth, and that immediate usefulness is neither the most successful nor the highest motive in scientific research, nor should research be limited by so narrow a condition. The investigations also made by the aid of Government Grants possess the usual degree of definiteness and of utility of such labours, and it cannot be reasonably expected that the world would be unanimous respecting any measure, especially respecting a subject so little understood by the public as the Endowment of Research.

If investigators were to limit their researches to utilities, or what appeared to be such, scarcely any essentially new experiments or new discoveries of importance would be made. No attempts would be made to discover essentially novel facts, nor would many trials be made to test fundamental abstract questions which affect the very basis of scientific knowledge. The principles of electro-magnetism, of magneto-electric action, and of the magnetic rotation of polarized light, were each discovered by means of perfectly novel experiments, in which immediate utility was not the motive.

It is worthy of notice that of the very small proportion of scientific investigators who disapprove of State aid to Research, nearly every one already possesses sufficient pecuniary means to carry on investigations, and therefore cannot adequately appreciate the position and necessities of investigators having only small incomes. In some cases also the objections to aid investigators come from scientific men who have attempted to make discoveries but have not succeeded.

Dr. Robinson of Armagh, a well-known investigator, has very properly pointed out * what has been done in this country towards giving assistance to those engaged in the pursuit of science, and mentions the Observatories maintained by the Universities and by the Nation. He says also that if anything more were to be done in increasing the amount of grants of money to assist scientific work, he thinks " it might be best applied in establishing in the great commercial

* *" English Mechanic,"* August 17th, 1881, p. 83.

centres of the realm, physical and chemical laboratories such as that which the Duke of Devonshire has
established at Cambridge, provided with the most
refined apparatus, and accessible to all who are considered privileged by a competent tribunal." He also
says "when there is found a man so far surpassing his
fellows in any department of science that he may be
expected to do work beyond their power, he ought to
be made independent of any other pursuit, so that
none of his time and energy may be lost, such a case
is exceptional, and when it occurs it should be excep-
tionally provided for."

Original research will for a long time to come, be
opposed by a large section of the non-scientific
public :—by the numerous persons whose source of
income depends upon the ignorance of their fellow-
men :—by those who are deficient of faith in demonstrable truth, and fear that their most cherished
beliefs are endangered by it :—and by many of those
who are insufficiently acquainted with it to perceive
its great value to mankind.

With regard to the fears of many objectors that the
Endowment of Research would lead to jobbery and
abuses, and thus retard the progress of discovery
instead of promoting it ; it is evident that such a risk
is an inseparable concomitant of every remunerated
office and is not peculiar to that of research, and must
therefore be accepted as unavoidable and be provided
against in the usual ways. It does not however appear
probable that the risk in this respect is at all greater

than that already existing and provided against in many other appointments.

Many persons, not clearly perceiving the difference between pure research and other scientific occupations, suppose that because science is encouraged in various ways in this country ; and because sums of money are occasionally given to scientific institutions, and some scientific men are evidently receiving good incomes, that discoverers are remunerated, but this is a great mistake ; there is probably not a scientific man in the kingdom who is wholly employed in such work in abstract physics or chemistry, and paid for his entire skill, time, and labour. Wherever payment is made for scientific labour, it is nearly always for that performed with a view to some immediate practical application. Inventors and expositors are remunerated, but discoverers are not.

At the present time in this country scientific men are paid for teaching, lecturing, writing popular scientific articles, compiling scientific books, editing scientific journals, making chemical analyses and experiments for manufacturers, companies, and others, for practical purposes, or to obtain evidence for legal cases, giving evidence on scientific subjects in courts of law, with consultations and advice to manufacturers and others, superintending scientific commercial undertakings, &c. Some also unfortunately obtain an income and cheap publicity by the empirical contrivance of selling to tradesmen, their scientific opinions in the form of testimonials which are extensively

advertised at the cost of the purchasers.   But not one
of these occupations constitutes pure research, or is an
immediate source of new discoveries.   Payment is
made for all kinds of scientific labour which will immediately benefit individuals or corporations, but
very little for pure investigation, and nearly every
inducement exists to attract men of science from
pursuing such labour.

It might be supposed that investigators would
patent or sell their discoveries ; but discoveries in
pure science cannot usually be patented or sold,
because they have not been converted by invention
into commercial commodities.   New scientific truth is
utterly unsaleable ; no one will purchase it.   Whilst
the real or intrinsic value of it is great, its extrinsic
value is small and is the sum of money it will sell for
in the market.   No one would have purchased Oersted's
great discovery of electro-magnetism.   It would also
be less to public advantage if investigators were to
neglect discovering new knowledge in order to apply
that knowledge to practical uses.   It requires a
different training of mental power to discover new
truths, than to utilize them by means of invention,
teaching or lecturing ; and men who can invent and
instruct are far more numerous than those who are
able to discover.   Discoveries are also generally much
more valuable than inventions, because a single discovery (that of gutta percha for example) not
unfrequently forms the basis of many inventions.
Discoverers not unfrequently meet with new facts

which they perceive might be applied to valuable technical uses, but they hesitate to patent them because the process of invention, taking out a patent, seeking a manufacturer to work it, and protecting their patent from piracy, would occupy a large portion of their time, and take them away from research. Sir D. Brewster got no money by patenting his kaleidoscope because the patent was instantly pirated in all directions.

Some persons have suggested that scientific men should keep their discoveries secret, but this would usually be a greater disadvantage to the investigator even than publishing them, and no one would then derive any benefit; discoveries also, being often capable of numerous applications, and not being in a saleable shape, cannot usually be monopolised by any one. New scientific knowledge is like a powerful light, it cannot be hidden. Discoveries are eminently national knowledge, and research should therefore be national employment.

Other persons suppose that investigators should be satisfied with the fame of their discoveries, and not require any payment; but this is a most unfair supposition, because no man can live without means, and every useful person deserves to be paid for his labour. Ought the late Duke of Wellington to have been satisfied with the fame alone of his exploits, without being paid any salary? Ought a Bishop to be content with the renown of his eloquence, without receiving any payment for his services? Genius alone is appropriately rewarded by fame, but time, unusual skill, labour and expenditure, should be repaid by money.

It has been suggested that an investigator, if he is a man of practical ability, is very often put into an office, the duties of which he can efficiently discharge, and yet have leisure for original research, as in the case of the late Dr. Graham, the eminent Master of the Mint,* our Astronomers Royal, &c., and thus obtain his reward. But this is a very imperfect plan, because research is very difficult, and to be carried out effectualy, requires the whole of a man's time and attention ; the investigator would also be taken from more important work to do that which is of less value to the nation, and which might be performed by a more suitable person ; appointments also of the kind referred to are much too few in number. Such a plan as this, of relegating important national work to odd hours spared from official duties, is a makeshift, and quite unworthy of this nation. Entire occupation in research, combined with efficient publication of the results, is the only satisfactory plan of procedure.

Probably one of the most satisfactory ways of rewarding scientific discoverers and serving national interests at the same time, would be to create salaried professorships of original research, and appoint discoverers of repute to fill them.

The time is near at hand when this nation will be compelled by the injurious consequences arising from its neglect of scientific research, to acquire a know-

* The Mastership of the Mint is no longer given to scientific men.

ledge of the relations of science to national existence
and welfare, and to adopt some means of encouraging
discovery. The greatest difficulty, probably, which
has to be overcome, is not so much how to obtain
funds for the purpose, as how to employ them successfully, and especially how to prevent their getting
into the hands of unsuitable persons. But, as methods
have been found of remunerating all other classes of
persons, ways may be devised of remunerating scientific investigators. It is only because the case is novel
that it seems difficult ; it is probably no more intrinsically difficult to establish a professorship of research
than to found an ecclesiastical benefice.

The great difficulty of determining from what source
discoverers should be paid for their labours, arises
from the fact that all classes of the community share
in the benefit. It is evident they should in some
measure be paid from a source towards which all
classes either directly or indirectly contribute, and
therefore from some public fund. The persons who first
use scientific knowledge are the compilers of scientific
books, and teachers of science ; but these only disseminate the knowledge, and do not derive from it
any great pecuniary advantage, they are only the
agents for supplying the knowledge to others. The
persons who first convert such knowledge into valuable commercial commodities are inventors and manufacturers who have received scientific education or
advice ; but those who derive the greatest pecuniary
benefit from it, and who should therefore either

directly or indirectly pay in the largest degree for it, are the great manufacturers, capitalists, and land-owners. Whilst the question is being settled as to what class of persons shall primarily bear the expense of research, discoverers themselves are suffering great injustice, and our manufactures and com-merce are passing into the hands of foreign nations. What the amount of loss and disadvantage suffered by this nation, through want of encouragement of scientific enquiry is, cannot be estimated, but it is certainly enormous. Had even a very moderate am-ount of payment been made for such labour, and the expenses out of pocket paid in full, the amount of research performed would have been greatly increased.

Under present circumstances, many promising young men, fitted to become good investigators, have been driven out of science althogether. I have found by long experience and persistent enquiry, that there are many young men distributed over this country, who are very desirous of engaging in scientific re-search, and likely to make good investigators, but are entirely prevented by the non-remunerative character of the labour , every one wishes to know " what will it lead to " ? Even amongst our most able discover-ers, scarcely one who has not possessed private means has continued research beyond the middle age of life, because such labour enables no provision to be made for old age ; and all those who have left have devoted themselves to less important but more lucrative occupations. Most of these gentlemen have been obliged to abandon research at a period of life when their faculties were in the most perfect state for continuing it.

Where one young beginner in science meets with the fortunate circumstance of a helping hand, as Scheele did in Bergmann, and Faraday in Davy, many are crushed out. The want of encouragement to scientific discovery in this country is so very great that extremely few men are able to struggle through it, and this is one reason why we have had so few discoveries. Some persons have argued that the very difficulties and discouragements offered are an advantage to science by producing only men of the very highest eminence in discovery; but it is manifest that however great the amount of ability may be that is developed by discouragement, that amount would probably be still greater by judicious assistance. Moreover, progress in the developement of the national scientific intellect is not so much to be reckoned by the few great successes which have occurred in spite of all obstacles, but rather by the much more numerous ones which would have resulted from proper encouragement. The advocates of such an argument can have no idea of the heart-sickening feeling of long deferred hope experienced by the young beginner in science ; or the disgust gradually engendered in his mind at the injustice of other men taking all the profits of his labours and leaving him without means of support ; or they would never adduce it. In this country the success of the few eminent men of science has resulted from the accidental combination of a few more or less fortuitous circumstances, and their own great natural determination, and not from legitimate and just support. How many investigators we have lost from the above causes it is impossible to tell. The encouragement also of unusual ability should not be left to accident.

As scientific research has proved itself to be of such great value to this nation, the question naturally arises, how can it best be promoted? A number of plans have been proposed. Amongst these may be mentioned. 1*st.* By founding State Laboratories, in which discoverers of established repute, supplied with every aid and appliance, should be wholly engaged in research in their respective subjects, and be paid by the State. 2*nd.* By founding colleges or Professorships of original research in each of the Universities, and appointing professors similarly. 3*rd.* By founding provincial colleges or Professorships of research, the funds being raised locally by means of subscriptions, donations, and endowments, with or without State assistance. 4*th.* By State or Local aid, in the form of additional salary, to Professors in colleges, to enable them to pursue research. 5*th.* By an extension of the present Government grants distributed by the Royal Society. 6*th.* By making it a condition at each of our Universities that every student entering for a degree in science, should previously make an original research. 7*th.* By the formation by learned societies, of Endowment of

Research Funds, and making grants of money therefrom to recognised investigators. *8th.* By aid to local scientific investigators by Municipal bodies out of the rates. And *9th.* The support of Institutes of Scientific Research by private munificence, Aid to research, in Germany, has chiefly been made by the State, by affording means to the Professors in the Universities; in America, more by munificence of wealthy individuals; and in this country, chiefly in the form of Government grants of money to investigators. The greatest difficulty to be surmounted in carrying out any of these schemes, is the very general ignorance in this country of the value and necessity of research; and this can only be overcome by scientific men themselves performing their duty of enlightening the public on the subject.

1st. By founding State Laboratories. One of the first duties of a Government is to protect its subjects in the enjoyment of their property ; but as no law reserves to discoverers the fruits of their ability, it is clearly a duty of the State to protect them in other ways. It is believed to be a duty of the State to provide and pay for pure scientific research, for the following reasons :—because research is eminently national work ; because the results of such labour are indispensable to national welfare and progress, and are of immense value to the nation, and especially to the Government ; because nearly the whole pecuniary benefit of it goes to the nation, and scarcely any to the discoverer ; because research is not sufficiently provided for by means of voluntary effort, nor can its benefits be restricted to a locality ; and because there appears to be scarcely any other way (except by application of University revenues) in which discoverers can be satisfactorily recompenced for their labour. Also " Government should for its own sake honour the men who do honour and service to the country." (Faraday.)

The founding of State laboratories for original research was proposed and advocated by the late Lieutenant-Colonel Strange, F.R.S., in communications read before the British Association, and in evidence given before the Royal Commissioners.* As the erection and maintenance of State laboratories would require a large sum of money, and as all classes of the community would share in the benefit, it is reasonable to suggest that the money should come from some source towards which all classes of the community, either directly or indirectly contribute, and therefore from some national fund.

In national improvements, expense is quite a secondary consideration ; the funds however for providing State laboratories already exist ; the sum of nearly £600,000 has accumulated in the form of fees received by Government for the granting of patents for inventions ; and as the discoveries made by scientific men form the materials by means of which those inventions were made, the money thus accumulated may be justly claimed by scientific discoverers as a suitable

See Reports of Royal Commission on Scientific Instruction and Advancement of Science, Vol. 2, pp. 75-92.

source from which their labours should be remuerated by the State.

Strong arguments might be adduced both against and in favour of the application of this money for the purpose. Inventors are a great wealth-producing class of the community ; they are at present very highly taxed, and receive but little advantage in return ; to tax them without giving them equivalent advantages, strikes very near to the root of commercial prosperity, by diminishing improvements in arts and manufactures. If an inventor is poor and his patent is valuable, he is also frequently harassed out of it by litigation. Inventors are usually poor, and but little able to pay taxes on patents at all ; their pecuniary position in this country is not greatly better than that of discoverers ; they are largely at the mercy of manufacturers and capitalists ; and the injustice to which they are sometimes subjected is notorious and disgraceful. On the other hand, the fund already exists ; inventors also receive an equivalent from investigators, discovery is the indispensable basis of nearly all invention ; patented inventions are formed by means of the knowledge obtained by pure scientific research. The poverty of a man does not justify his taking the fruits of the labours of another man without paying for them. If also the patent fees were thus applied, the cost of research would then be paid for by all classes of the community, somewhat in proportion to the benefit derived, because the cost of patents in general ultimately falls upon the public at large, in the shape of an increased price put upon the commodity, in order to pay the cost of the patent, like the grower of wheat is paid by the consumers of bread, through the medium of the baker, the miller, and corn merchant.

Whilst the benefits derived from the labours of discoverers, flows chiefly in the form of money into the hands of wealthy manufacturers, and finally gets locked up in the possession of capitalists and land-owners, it is hardly to be expected that the Government will be in possession of funds necessary to promote research unless some such plan as this is adopted.   Should the wealthy and governing classes however become sufficiently acquainted with the value of research, and of the essential and permanent dependence of their material prosperity upon it, there will then be some hope that they will be willing to contribute in a more direct manner their just share towards paying the expenses.   There is, however, but little prospect of this whilst the influence of wealth so depresses the scientific education of those classes at the Universities.

The fundamental object in founding State labora-tories should be to keep a staff of the most competent men wholly engaged in original research in pure science, and a secondary object might be to train as-sistants to become investigators.   Such laboratories would doubtless be located in London, and be on a scale of magnitude creditable to science and the nation.   They might very suitably include departments for the simpler pure sciences and even for biology.

It is manifest that the arguments which support the proposition for professorships of original research in those sciences apply in a greater or less degree to other sciences ; and it has been stated that " there is no ground upon which the scheme can be limited to the subject of natural philosophy." In reference to this remark, (which, like most objections, contains some truth), it must be remembered that there is a natural order of dependence of the sciences upon each other, in which order also they are being evolved. It is a general truth that the physical sciences of light and heat are based upon mechanical conditions of the particles of matter ; that the science of chemistry is founded upon physics ; that biological phenomena are dependent upon physical and chemical conditions ; that psychological subjects are based upon biology ; and that biological science cannot progress excepting in proportion to the advance of the sciences on which it is based. In addition to this general order of pre-cedence of evolution in point of time, the various sciences are so mutually related that all must advance together, the simple ones taking the lead, and the concrete and more complex sciences, with their at-tendant arts, following behind.

As this natural order of dependence and develop-ment of the sciences is a great fact of nature, over which we have little or no control, and as a scheme for the simultaneous establishment of professorships of research in all the sciences, simple, complex, and concrete, would probably be too great an undertaking, the most proper course would be to commence with the simpler ones, such as those of mathematics, mechanics, physics, and chemistry, and perhaps biology. If it be argued that it would be unadvisable to commence with the simple and purely experimental sciences, it would be still more unadvisable to commence with the concrete subjects of "natural history, medicine, civil history, law, and theology," or with the arts which also depend upon science.

The number of investigators in such an institution would not be large, because few of high repute could be obtained, many of our ablest ones abandon research for remunerative pursuits. In order to make the plan succeed, the conditions of the appointments should be such as to limit the election to the most competent persons. In the selection of such gentlemen, the verdict of opinion of scientific men generally, upon the published researches of the candidates, would have previously determined who were qualified for the office. Any man who had published reliable papers in the Transactions of the Royal Society, might very properly be considered a fit candidate, and the selection and appointment might be made by the Government, with the advice of the Council of the Royal Society.

Probably there exists no class of persons upon whom the country might more rely for industry in office than eminent investigators, because they have pursued truth for its good effects alone. Men who had

previously exercised the degree of self-sacrifice necessary to make a number of long and difficult experimental researches, with only very limited pecuniary
means, must necessarily be possessed of great enthusiasm in their calling, and would therefore be
extremely unlikely persons to become idle by being
supplied with a sufficiency only of means to carry on
their labours. Further, such men might at present
obtain a much larger income than they wonld receive
in such a post, by abandoning research and devoting
themselves to the various profitable engagements
which are open to every man of scientific ability who
is willing to devote himself to applied science. The
actual work of research is much too arduous and difficult to permit such an office to become an object of
desire to a place-seeking or idle person. But in order
to exclude with certainty those who might devote
themselves to research solely or primarily for the purpose of subsequently obtaining a well paid appointment, (as persons sometimes devote themselves to learning, with the object of getting an "idle Fellowship,")
and to ensure in all cases a reasonable continuance of
industry, it would be necessary, that whilst the salaries
paid should be sufficient to render the professors free
from care, if expended with a reasonable degree of
economy, they should not be so large as to conduce to
idleness. The professors should undertake not to
engage in any other remunerative employment, and
provision should be made, that in case a professor
persistently failed to make, complete, or publish his

researches, or devoted less than the stipulated amount of time to such labour in the Institution, without reasonable cause, he should be removed.

Many persons fancy that "it must be very nice to be always making experiments," and that they "should be delighted with such an occupation" if they "could only spare the time." But such an idea is only another illustration of the general ignorance of the subject, and it is only expressed by those who have never made a laborious and difficult research. Pure research is by far the most difficult of all scientific occupations, and this is another chief reason why discoverers are few, and why they will probably remain so.

To succeed in research, a man must set aside all human pride, and approach the subject with perfect humility ; and this is not an easy task, men cannot so readily abandon preconceived and cherished notions. Many researches are moreover extremely dangerous. Thilorier was killed by the explosion of a vessel of liquefied carbonic anhydride ; Dulong lost an eye and finger, and Sir Humphrey Davy was wounded by an explosion of chloride of nitrogen. Faraday was near being blinded by an experiment with oxygen. Nicklès of Nancy, and Louyet of Brussels, lost their lives, and two other chemists were seriously injured in health by exposure to the excessively dangerous fumes of hydrofluoric acid. Bunsen lost the sight of an eye and was nearly poisoned by an explosion whilst analysing cyanide of cacodyl.*   Hennel was killed by an

* See " *Nature* " No. 603, p. 597, April 28th, 1881.

explosion of fulminate of silver, and Chapman by one of nitrate of methyl ; and nearly every chemical investigator could tell of some narrow escape of life in his own experience. Any one who wishes to know whether it is "very nice to be always making experiments" should attempt the isolation of fluorine, the chemical examination of some offensive substance, or the determination of some difficult physical, or chemical problem.

That a professorship of original research would "involve substantial work" does not admit of doubt, and therefore "there would be some security that it would be worthily bestowed." It would not become an "ornamental sinecure," in which "there is pay but no work," unless, by assigning to it too large a stipend, inducement was held out to that numerous class of persons whose love of money is stronger than their love of truth, to seek the office ; to say the utmost, it could hardly become so largely a sinecure as many offices now held by ecclesiastics. Jobbery and abuse of patronage would be still further prevented by making the duties sufficiently heavy.

The appointment, and remuneration by salary, of professor of research, would not lessen the independence of scientific men if the office was not placed under the superintendence of active and interfering officials ignorant of science. Although the professors might not be highly paid, the appointment would increase their independence. because it would be one of the most honourable to which scientific men could attain, and because they would thereby be put into a sphere in which they could exercise their talents to the fullest extent, and render the greatest service and honour to the nation. If also the salaries offered were not too great, those persons only would become candidates who at present have insufficient means to defray the considerable cost of experiments.

It would be necessary to appoint only persons who would undertake to devote their time solely to the discovery of new facts and principles in science, and the determination of purely scientific questions, and not to the making of inventions; because discovery is of far greater national value than invention; and because inventions would immediately on their publication be seized, modified, and patented by individuals for their own personal benefit. Discoveries, on the other hand, would require a large additional amount of labour expended upon them by inventors before they could be converted into saleable commodities.

Each professor should be allowed perfect freedom to choose his own special subjects of research in the sciences he had been accustomed to study, because each investigator is usually the best judge of what researches are the most likely to yield him important results. No discoverer of repute would be very likely to trespass on another man's sphere of research, because he would usually have an abundance of good subjects of his own ; and every honourable man would purposely avoid doing so ; and we find this practically to be the case at the present time. Separate sets of rooms would be necessary for each investigator in order to keep the researches private and distinct.

The whole of the new knowledge obtained by research should be treated as national property, and all of it worthy of publication should be made known without the least reserve, it would also be desirable to publish the results at reasonable intervals of time. The publication might take place, as at present, in the journals of the learned Societies, or in the leading scientific magazines, and the value of the work would be largely guaranteed by such a mode of publication· The professors should also engage not to sell, patent, or prematurely disclose any of the knowledge obtained. By electing to such offices only discoverers of repute, the nation might reasonably depend for the results upon the known ability and industry of the men. That the results obtained would, many of them, be highly valuable, does not admit of doubt, because long experience has uniformly proved it ; but no discoverer can tell beforehand what results he will obtain, otherwise research would hardly be needed.

An objection has been made that no one can tell how long it will be after a discovery is made before the nation will derive the chief benefit. The length of time which elapses between the publication of discoveries and their practical fruits is very variable. Usually benefit commences at once and gradually widens ; directly discoveries are published they begin to be used by compilers of scientific books, and by teachers and lecturers in science, and are thus diffused amongst the public in general, and begin to produce beneficial effects.   Inventors, manufacturers, medical men, and others, also begin to apply them to their respective purposes.   In some cases striking applications are immediately made of them, and public attention is thus directed to the useful result ; but in many cases the beneficial effects are small, numerous, and indirect, and it is difficult to trace and describe them. The objection also is deficient in force, because expenditure in any other occupation, and receipt of the profit upon it, are rarely simultaneous.   Many of the wisest reforms in this country have been a long time in producing their results.   We must therefore be content, as in all ordinary cases of investment, with the conviction that the expenditure will be profitable. and we must wait patiently for the certain harvest.   In research, as in many other human enterprises, a man who will not move until he is absolutely certain that what he intends to do will at at once succeed, must sit still and perish.

Suggestions have also been made to appoint a Government Committee, or Council, whose function should be to value scientific discoveries, and make corresponding amounts of reward to the discoverers. But this appears to be a less feasible plan, because no man can, at the period of discovery, determine what amount of practical result a discovery will ultimately produce.   Who could have foretold with certainty at the date of Oersted's discovery of electro-magnetism, that this discovery would result in the expenditure of hundreds of millions of pounds upon telegraphs alone ? *

Objections have been made to definite payment for labour in research, on the ground of indefiniteness of the results, and the impossibility of measuring their value. Can we expect to buy new scientific knowledge at so much a pound, or to retail discovery by the pint ? The work of discoverers is as definite as that of many other persons who are paid. Who can measure the value of the cure of souls, of the duties of a judge, or of those of a field-marshal ? Instead of paying for the labour of research in a definite way, we have adopted unsatisfactory makeshifts. Exceptional gifts, and semi-charitable pensions, have been with difficulty obtained in a few cases for scientific men ; most often for those who applied scientific knowledge to practical uses than for those who discovered that knowledge. In this country, neither lawyers, medical men, military persons, nor clergymen are paid definitely by results, but by time and labour, in accordance with the re-putation of the man, and there is no sufficient reason why investigators should not be similarily remunerated. The differences in the cases are only ones of degree.

The time has arrived when this great evil should be made known and remedied, and men of science should press upon our Government, as a matter of justice to themselves, and necessary for the nation's welfare, that the accumulated fees from patents should be applied to the establishment of a Scientific depart-

---

* A fleet of thirty ships, varying in size from 500 to 5000 tons each, is employed in laying and repairing telegraph cables, and 25 millions of pounds have already been invested in submarine cable enterprises.

ment of the State, the erection of State laboratories, and the payment of discoverers for the national work of research.

2nd. Professorships of Research at the Universities. Most of the remarks already made respecting the appointment and maintenance of professors of research in State laboratories, would apply equally to those proposed in connection with the Universities. Amongst the reasons which may be adduced in favour of the establishment of such professorships the following may be selected. Because the advancement of learning is peculiarly the function of a University, and one of the chief objects for which such institutions were founded. The word University implies a seat of universal knowledge, and it is reasonable to assume that Universities should act as fountains of new theoretical knowledge, as well as perform the function of diffusing it; such an addition would also raise their intellectual position, and make them much more respected, both at home and abroad.

With regard to such professorships, the "Association for the Organization of Academical Study," consisting of a number of learned men belonging to the Universities, the Royal Society, and other learned bodies, adopted the following resolutions :—

" That the chief end to be kept in view in any redistribution of the revenues of Oxford and Cambridge is the adequate maintenance of Mature Study and Scientific Research, as well for their own sake as with the view of bringing the higher education within the reach of all who are desirous of profiting by it." " That to have a class of men whose lives are devoted to research is a national object." "That it is desirable, in the interest of national progress and education, that Professorships and special institutions shall be founded in the Universities for the promotion of Scientific Research." " That the present mode of awarding Fellowships as prizes, has been unsuccessful as a means of promoting Mature Study and Original Research, and that it is therefore desirable that it should be discontinued."

With regard to the funds necessary :—It has been estimated that the money paid in the form of sinecure fellowships or retiring pensions, to young men in Oxford alone "now amounts to about £80,000 or £90,000 a year ; and it has been suggested that this money be applied to the purpose. These funds were originally intended for promoting knowledge, but vested interests prevent their being used for discovering new truths.

The chief object of such professorships would be the same as that in the proposed State laboratories, viz.— to keep a staff of the most competent men wholly engaged upon research in pure science. The professors of physical and chemical research might be selected in accordance with the suggestions already made, and be appointed by the Senate or other governing body, with the advice of the Council of the Royal Society. All the precautions which have been already suggested under the head of " State laboratories," would have to be taken in order to exclude unsuitable persons, and to secure industry in the professors. The remarks also, already made respecting the limitation of the duties of the professors to research in pure science, the exclusion of invention, the publication of results, the class of sciences with which a commencement might best be made, etc., apply equally in this case. I do not however mean by these remarks to suggest the disendowment of research in the more complex or concrete subjects, in order to make a commencement with the simpler sciences.

The existence within the Universities of offices in which the faculties of scientific men might be developed to their fullest extent, would induce those engaged in the work of scientific instruction in those institutions to devote more time to research, in order that they might improve their scientific talents, and in their turn become fitted to occupy such posts.

It has been suggested that discoverers should teach as well as investigate ; but this would be an imperfect plan, and would largely convert the position of a professor into that of one at an ordinary college. Every person who has had much experience in experimental investigation also knows, that to carry it out effectually, requires the whole of his time and attention. If, therefore, teaching or lecturing constitutes a part of the duties, a portion of the professor's time must be taken from the more important occupation of research, and the fundamental object of the institution will be frustrated. Research evolves new knowledge ; teaching simply distributes it. The labours of a scientific teacher or lecturer consist esssentially of a continued series of repetitions of other men's discoveries. For each single man who can discover, there exist many who can teach. With teaching in addition to research, all the usual educational machinery, lecture theatres and apparatus, diagrams, audiences, pupils, registration of students, receipt of fees, examinations, marking of papers, valuing of answers, attending annual meetings, etc., would be brought into requisition, and the result would probably be, as it is at present, the duties of teaching would, in nearly all cases, swallow up the time, and prevent the freedom from interruption necessary for successful research Under present circumstances, it is the testimony of nearly every teacher and lecturer in science, that he " has no time for research." If teaching is also carried on, the Research laboratories will compete with educational institutions, established and carried on by private enterprize, and place them at a disadvantage, and thus discourage voluntary effort in the diffusion of science ; but by limiting the functions of the professors entirely to pure research, there will be no competition with any private interests, because no persons gain a livelihood entirely by means of such occupation.

That the practice of teaching is however of very great use in preparing the mind of a scientific man for research is quite certain, because it compels him to study all parts of his subject, and whilst doing so, many questions for investigation occur to his mind.

Many of our most eminent discoverers have also been teachers. But teaching is not a necessary condition of success in research, nor even a necessary preparation for it ; the examples of various able discoverers prove this. In some cases discoverers have devoted themselves to teaching, only after having attained high repute in research ; in others they have not been teachers at all. Original research itself usually suggests plenty of new subjects of investigation, without the additional ones suggested by teaching.

In order that self improvement in a man of science might never stagnate, there should exist a continuous and complete series of steps of preferment, by which the merest beginners in scientific knowledge, might be enabled to attain to the highest scientific position ; and finally become wholly occupied in that kind of labour by which their scientific faculties would be developed to their fullest extent, but this last step is wanting. By the proposed plan however a student would become a teacher ; the teacher develope into a professor ; and a professor might employ for a period a portion of his time in research, and thus become qualified for entire devotion to original investigation and discovery.

After a scientific teacher has acquired a thorough knowledge of his subjects, and a high position in his professsion, his occupation becomes to him a species of intellectual routine, in which he is continually going through the same courses of lectures and examinations over and over again, and his personal improvement stagnates. But if there was remuneration for research, or there existed some post or employment, to which those who had acquired the ability to investigate might be appointed, there would be an inducement to continued intellectual improvement, and a sphere in which the most valuable faculties of scientific men might be developed for national benefit to their fullest extent.

3rd. Provincial Colleges of Research.—The success of this plan would depend essentially upon the diffusion of a knowledge of the importance of scientific research amongst the richer classes. There are at present a very few wealthy persons in this country who perceive to some extent the value of such research, and the dependence of their wealth upon it, who would be willing to contribute to a fund for the purpose ; and there are many more who would assist if the importance of the subject was properly explained to them. The chief argument in favour of provincial colleges of research is, that it is a duty of wealthy persons to aid research, because they have derived, and are continually deriving great benefits from it, for which they make no payment. The ways in which some of those benefits have been derived have been already briefly stated. As the large manufacturers and capitalists are generally the persons who derive great pecuniary benefit from the progress of science, it might be reasonably urged that they should contribute freely towards its advancement.

Such an institution might be located in each of the

largest centres of industry. The objects of the institution; the branches of science to be investigated in it; the number of professors, the mode of selecting them, and of excluding unsuitable candidates for the office; the means by which industry might be secured and jobbery prevented; the exclusion of invention, and of teaching and lecturing; the publication of results, removal of professors, etc., have already been treated of under the head of "State laboratories." The chief difficulties to be overcome in this, as in all other plans of aiding research, are to find a sufficient number of influential persons acquainted with the subject to practically carry out the plan; to secure investigators of high ability; and to prevent the offices being filled by incompetent persons.

4th. Aid to Professors of Science in Colleges.—Another way by which research might be promoted, would be by giving assistance in the form of a definite amount of additional salary, for the purpose of pure research, to professors and teachers in colleges and institutions; the money being supplied by the State or from the funds of the Institution. In carrying out this plan, it would be necessary to assist only those persons who had already published a good research, and thus proved their ability; and who would engage to devote a definite portion of their time to the labour as a part of their duty. The selection of suitable men might be made with the advice of the Council of the Royal Society.

The additional salary should be entirely in the form of remuneration for labour, time, and materials, etc., expended upon research in pure science, and not in effecting inventions. The knowledge obtained should be treated as public property, and be published in the usual manner, and the investigator should not be permitted to sell or patent it. It would be necessary to provide that in case the investigator failed to make or publish a reasonable amount of good research, the additional salary should cease. Publication in the journals of the Royal Society, or in a leading scientific magazine, might be considered a sufficient proof of the satisfactory quality of the labour.

It is very desirable that all the higher teachers and professors of science in our educational institution should devote a portion of their time to original research. It would make their lectures more reliable, because research yields experience in the detection of error; whilst there is usually only one way of succeeding in making an experiment, there are always many ways of failing, and in the directions given in books, the latter are usually omitted. It would also induce the students to take a greater interest in the subject, and feel more respect for the teacher. The special excellence of the German system of teaching consists in the union of teaching and original research. This plan of aiding research would induce some of our teachers of science who have not yet made researches, to attempt such labour, it would also develope a superior class of scientific teachers generally; and produce a supply of candidates for professorships of research.

A great obstacle to the carrying out of this plan lies in the fact that in consequence of the ignorance of the value of original research by the founders of such Institutions, no definite provision usually exists in the Trust deeds to authorise the Trustees to devote any of the funds to such a purpose.

5*th. Extension of the Government Grant System.*— During a number of years the British Government has entrusted to the Royal Society the annual sum of £1,000 for the purpose of aiding science; and that sum has been given in varying portions to different investigators who have applied for grants in aid of their expenses in making investigations.

Although the total amount to be disbursed annually was not large, very few persons, qualified to make good researches, usually applied for its assistance, and it was difficult to dispose of the whole. The chief causes of this difficulty were :—a grant from the fund was an unprofitable gift to accept, because it was only sufficient to partly pay the expenses out of pocket for chemicals and apparatus, and allowed nothing for the skill, time, or labour, nor for payments made to assistants. Further, " By order of the Council, all instruments, apparatus, and drawings, made or obtained by aid of the Government Grants, shall, after serving the purpose for which they were procured, and in the absence of any specific understanding to the contrary, be delivered into the custody of the Royal Society."

By far the greater part of the expense of an investigation in physics or chemistry is the exceedingly large amount of time it occupies. Many necessary preliminary experiments have to be made, which yield either negative, unsuccessful, or incomplete results, and make the undertaking expensive. A good investigation in chemistry also not unfrequently costs the investigator a sovereign a day if he is wholly employed upon it. In some cases, for each £100 received as a grant, at least a £1,000, was directly and indirectly expended. Any person therefore who undertook a research became a loser, and aid from the Government Grant fund did not entirely cover his loss. Only scientific men who had other sources of income were able to avail themselves of the grants. The existence of the grants also was not widely known. The advantages of the plan were, it diminished the loss to the investigator, and the fact of being allotted a sum from the fund was considered highly creditable to the recipient.

In consequence largely of the evidence collected from eminent men of science from all parts of Great Britain, and the recommendations based upon it, by the Royal Commission for the Advancement of Science, the Grant system has been extended ; our Government recently placed an additional amount of £4,000 a year, for five years, to be distributed in sums at the recommendation of the Royal Society to suitable applicants, and the five years have now elapsed.

This extension of the Grant system has been an

improvement. It has resulted both in a large increase in the number of applicants and of researches ; and has shown that there exists in this country a large amount of scientific ability in need of encouragement. The amounts granted were increased in magnitude so as to cover in some cases payments made to assistants and the entire outlay made for experiments, also a small payment for a portion of the time occupied in actual research. The plan of awarding the grants has been for work proposed by the applicants to be done, and not for that already performed. How far a retrospective method, might be worthy of trial, is difficult to decide. It has been objected to it that the claims of scientific investigators for researches already made, would be so great and so convincing that it would be impossible to resist them, and the funds required to satisfy those claims would be so large as to render the plan quite impracticable ; if however the retrospective period was limited to a short time, a year for example, the difficulty would be lessened. There would still however remain the great difficulty of valuing the results. This might probably be overcome by regulating the money payment according to the time, labour, pecuniary expenditure, and scientific status of the particular investigator, and leaving genius to be rewarded by the fame and honour of the results.

No system of aid however can place scientific investigation in a satisfactory position in this country, which does not include certain remuneration for time, money and labour expended; and no sound argument can be adduced why investigators should not be adequately recompenced. The genius alone of a discoverer should be rewarded by fame, and his time, labour, and expenditure, in accordance with his professional reputation, be repaid by money, as in all other intellectual occupations. The same amount of time and labour expended in any ordinary profession, requiring an equal, or even less amount of preparatory education and experience, and less rare ability, would yield an income of several thousand pounds a year. Although the lives of a few eminent discoverers have proved that it has been possible for them to do a considerable amount of research under the conditions which have existed, that is no reason why they should not be remunerated. Previous success in research has been due to the unusually great perseverance, industry, and self-denial of the men, and but little to any pecuniary encouragement received. The fewness of such men, supports this view of the case. The plan of aiding research by grants which include no certain payment for time or labour, is quite incommensurate with the importance of the subject and entirely unworthy of the reputation of a great nation.

6th. Students pursuing Research at the Universities. In the German Universities each student is required to make an original research before he can obtain a degree in Science, and the plan has worked successfully; also in the Victoria University, Manchester, several Fellowships have recently been established for the encouragement of students in original investigation.

If this plan could be carried out in our old Universities it would produce most valuable results, because the governing, wealthy, and influential classes of this nation are chiefly educated at those institutions, and they would then acquire habits of more accurate scientific thought, and some knowledge of the nature and importance of scientific research, and of the essential dependence of national welfare upon it.

But a great and probably insuperable obstacle exists to the carrying out of such a plan, viz., the wealth possessed by the parents of students.   An original research cannot be made without considerable industry, and the greatest opponent of industry, especially with young men, is the possession or expectation of wealth. According to college tutors at our old Universities, there is no large class of industrious students at those institutions.   The greatest cause of the idleness of the students is parental neglect and the habits of wealthy society.   Many parents allow their sons too much money, and over-look too readily their idleness and frivolity ; the young men also know their parents are rich, and act accordingly.   Many persons send their sons to those places chiefly to form aristocratic acquaintances, and for other purposes than those of educational discipline and learning.   The college authorities have also largely acquiesced in the wishes of the parents and students.   And in this way scientific research has been almost entirely excluded from our old Universities, If the present tutors and governing bodies of those Institutions cannot in- duce students generally to be industrious, by what means can it be expected that these young men can be persuaded to exercise the still greater degree of industry and intelligence requisite to prosecute re- search, whilst they are decoyed from it by the attrac- tions of wealth ? In Germany the conditions are very different, the students in the Universities of that country have much less money at their disposal. Nearly the whole of the educational courses also at the Grammar schools and other educational insti- tutions in this country, are formed upon the plan of sending all the superior scholars to our Universities, and thus the defective state of scientific training at the Universities operates through our whole scholastic system, and depresses the entire scientific instruction of the nation. It is evident that in this way the undue wealth of this country largely retards national progress.

7th. Local Endowment of Research Funds. In addition to the foregoing means, local efforts might be made to encourage research in each great centre of industry ; through the medium of the local scientific societies. Nearly as early as the year 1660, Cowley in a treatise, proposed a Philosophical Society to be established near London, with liberal salaries to learned men to make experiments ; but he could not get the money raised. A plan of this kind is in operation in Birmingham and carried out by the

Council of the Birmingham Philosophical Society in accordance with the following :—

"SCHEME FOR ESTABLISHING AND ADMINISTERING

A FUND FOR THE

ENDOWMENT OF RESEARCH IN BIRMINGHAM"

"The Council are of opinion that this Society would be omitting a principal means of the advancement of Science—the end for which all such associations exist—if it neglected the question of the Endowment of Research. To maintain a successful investigator in his labours, even though no results of immediate or obvious utility can be shown to spring out of them, is of interest to the community at large. Indeed, it is just because the practical usefulness of such work is not immediate or obvious that it becomes necessary to give it special support, for otherwise it would have its own market value, and endowment would be superfluous. But the proper and effectual administration of an Endowment Fund is perceived to be so beset with difficulty, as often to deter even those who recognise the principle from advocating it in practice. Most of the dangers usually foreseen would, however, as a rule be avoided, simply by the distribution of such funds from local centres, under such a scheme as is now proposed. The Council, are therefore, anxious to establish a Fund, in connection at once with the Society and the Town, for the direct Endowment of Scientific Research."*

* See "*Nature*," vol. XXII, page 203.

8*th. Local Laboratories of Research.* Another plan
would be for local scientific societies to raise money
by soliciting subscriptions and donations for the support of local laboratories ; a prospectus of the following kind being issued :—

PROPOSAL TO FOUND A LABORATORY OF PURE

SCIENTIFIC RESEARCH IN ——.

"As the manufacturers, merchants, capitalists, land-
owners, and the public generally, of this town and
district, have derived and are still deriving great
pecuniary and other benefits from the discovery of
new knowledge by means of pure research in the
sciences of Physics and Chemistry ; and as in consequence of the great neglect of such research in this
country, and the increased cultivation of it in other
lands, our commerce is suffering, and a great many
evils in manufacturing and other operations, in sanitary and many other matters dependant upon physical
and chemical conditions, remain unremedied ; it is
proposed to found a Local Laboratory of original
research in those sciences, with every suitable appliance in it; and to employ one or more investigators
of repute, with assistants, who shall be wholly engaged
in such labour in their respective sciences."

As it is largely the custom in this country to effect
great objects by means of individual liberality and
corporate enterprise, instead of trusting to State
assistance, it is not improbable that when the
great importance of scientific research and its claims

to encouragement have become more generally known, that aid which has hitherto been with-held from it will be rendered by private generosity ; and local institutions, wholly for the purpose of original scientific research will be established and supported by public-spirited wealthy persons. An institution of this kind upon a small scale, and called "The Institute of Scientific Research" has already been established in Birmingham, (see Note p. 40). By founding local institutions of this kind there exist opportunities for wealthy persons to do great good to mankind, and acquire renown as philanthropists by the action.

And 9*th*. In consequence of the great benefit derived from scientific research by the inhabitants of each locality, it has become a duty of each large community to promote it, and local Town Councils might with advantage and perfect justice to the public, devote a portion of municipal funds to the purpose of aiding local scientific research. To this plan it may be objected, that as the results of research are cosmopolitan, diffusing themselves everywhere, and this diffusion cannot be prevented ; the benefits arising from research cannot be restricted even to a large community. In reply to this :—As knowledge and its advantages are cosmopolitan, the duty of promoting research must be equally extensive. There is also a real return received by the public for expenditure of money in research, in the free liberty to use all new knowledge developed everywhere by such labour, and although the money expended by a community upon particular researches or upon an individual investigator, does not directly produce an immediate return ; practically an immediate and direct benefit is received by that community, because new scientific knowledge for the use of teachers and popular lecturers, and new inventions based upon it, of of local value to that society, continually become public. Every civilized community has also received beforehand such benefits to an enormous extent ; and each investigator may reasonably claim public support on the ground that he contributes to the general stock of new knowledge. Some persons however, who have not fully considered the subject, wish to receive not only the advantages accruing from the common stock of knowledge, but also to reserve to themselves the entire benefit arising from their own special contributions.

Experience alone will prove which of the foregoing schemes is the most suitable in this country, or in particular cases. At present the plan largest in operation is the system of Government Grants, next in magnitude are the other funds distributed by the Royal Society, the British Association, the Chemical Society, the Royal Institution, the Birmingham Philosophical Society, and those provided by the munificence of private individuals. It is greatly to be hoped that the liberal spirit of private individuals will yet further remove the great blot which lies upon the reputation of the wealthy manufacturers, capitalists, and land-owners, who have derived such great profits from scientific research and have scarcely aided it at all in return. It is also to be desired that the Corporations of manufacturing towns will recognise the value of original scientific enquiry to their fellow townsmen, and will undertake the responsibility of voting money from municipal funds to promote it.

INDEX